SOLUTIONS MANUAL

to accompany

INTRODUCTION TO RANDOM SIGNAL ANALYSIS AND KALMAN FILTERING

ROBERT GROVER BROWN
Iowa State University

JOHN WILEY & SONS
New York Chichester Brisbane Toronto Singapore

ISBN 0-471-88052-3
Printed in the United States of America

10 9 8 7 6 5 4 3 2 1

CHAPTER 1

1.1 Total no. of possible hands $= \binom{52}{5}$

$$= \frac{52!}{5! \; 47!} = 2,598,960$$

Number of Heart flushes $= \binom{13}{5}$

$$= \frac{13!}{5! \; 8!} = 1287$$

Number of Spade flushes $= 1287$

Same for Clubs and Diamonds

∴ Total possible flushes $= 4 \cdot 1287$

∴ $P(Flush) = \dfrac{4 \cdot 1287}{2,598,960} \approx \dfrac{1}{500}$

1.2 Total no. of Black Jack deals

$$= \binom{52}{2} = \frac{52!}{2! \; 50!} = \frac{52 \cdot 51}{2} = 26 \cdot 51$$

Total Black Jack combinations of ace
and face card:

A of Spades ⟨
K,K,K,K (any suit)
Q,Q,Q,Q (any suit)
J,J,J,J (any suit)
10, 10, 10, 10 (any suit)

16 total

A of Hearts ⟨ etc.

∴ $P\begin{bmatrix} \text{Any Ace and} \\ \text{K, Q, J, or 10} \end{bmatrix} = \dfrac{4 \cdot 16}{26 \cdot 51} = \dfrac{32}{663} \approx \dfrac{1}{20}$

1

1.3 LABEL BALLS W1, W2, B, R1, R2, R3, and R4

Total No. of Outcomes $= \binom{7}{3} = \frac{7!}{3!4!} = 35$

(a) 35 elements (outcomes) in sample space.

For part (b) use a direct itemization approach:

$$W1 \begin{cases} R1, R2 \\ R1, R3 \\ etc \end{cases} \quad \begin{matrix} 4 \text{ things taken 2} \\ \text{at a time} = 6 \end{matrix}$$

$$W2 \begin{cases} R1, R2 \\ R1, R3 \\ etc \end{cases} \quad \text{Same as above}$$

∴ P[One white and two red] $= \frac{12}{35}$

1.4
(a) Sample space consists of 6^5 elements.

Examples: $\left. \begin{matrix} 1 1 1 1 1 \\ 1 1 1 1 2 \\ \vdots \\ etc. \\ 6 6 6 6 6 \end{matrix} \right\}$ $6^5 = 7776$ total elements

(b) Possible ways of exactly one six:
6xxxx, x6xxx, xx6xx, xxx6x, xxxx6
where x is any number other than 6.

∴ P[Exactly one six] $= (5)\left(\frac{1}{6}\right)\left(\frac{5}{6}\right)^4 = \left(\frac{5}{6}\right)^5$

$= \frac{3125}{7776} \approx .4$

(c) P[No sixes] $= \left(\frac{5}{6}\right)^5$

∴ P[At least one six] $= 1 - \left(\frac{5}{6}\right)^5$

2

<u>1.5</u> The probability assignment on the sample space is only constrained by the axioms of probability. However, it is reasonable to assume that the track odds on the various horses are adjusted so that each bettor stays even <u>on the average</u> (or loses the same percentage when the track "take" is accounted for). Therefore, imagine betting 1 unit on Horse #1. Say the horse wins, and you then have 3 units. The horse must then lose 2 races to bring you back even if you continue betting on horse #1. Clearly, Horse #1 must then win 1 out of 3 races <u>on the average</u> to keep its bettor even. A similar argument applies to the other horses. Therefore, arbitrarily assign probabilities in this case as follows:

$$P[\text{Horse \# 1 wins}] = \frac{1}{2+1} = \frac{1}{3}$$

$$P[\text{Horse \# 2 wins}] = \frac{1}{3+1} = \frac{1}{4}$$

$$P[\quad \text{''} \quad \# 3 \quad \text{''} \quad] = \frac{1}{5+1} = \frac{1}{6}$$

$$P[\quad \text{''} \quad \# 4 \quad \text{''} \quad] = \frac{1}{5+1} = \frac{1}{6}$$

$$P[\quad \text{''} \quad \# 5 \quad \text{''} \quad] = \frac{1}{11+1} = \frac{1}{12}$$

(Note that the probabilities sum to unity.)

3

1.6

Think of 3 discs of equal thicknesses. The thickness of overlapped regions is noted on figure.

$$P(A \cup B \cup C) = \underbrace{P(A) + P(B) + P(C)}_{\text{Includes overlaps}}$$

$$\underbrace{- P(A \cap B) - P(B \cap C) - P(A \cap C)}_{\substack{\text{Subtracted to account} \\ \text{for double overlaps}}}$$

$$+ \underbrace{P(A \cap B \cap C)}_{}$$

Previous terms "over compensated" for the triple overlap region so one "thicknes" must be added back.

1.7 Imagine 27 specific cards exposed to the declarer including 11 specific trumps. This leaves 2 trumps outstanding that will be denoted T1 and T2. The possible opposing hands may be categorized as follows:

1.7 (con't.)

Left Opponent		Right Opponent	
(a) T1, T2, $\times\times\times\cdots\times$	(10 x's)	$\times\times,\times,\cdots\times$	(13 x's)
(b) T1, $\times,\times,\times\cdots\times$	(11 x's)	T2, $\times\times,\times\cdots\times$	(12 x's)
(c) T2, $\times,\times,\times\cdots\times$	(11 x's)	T1, $\times\times,\times,\cdots\times$	(12 x's)
(d) $\times,\times\times,\cdots\times$	(12 x's)	T1, T2, $\times\times,\cdots\times$	(11 x's)

Calculation of number of possibilities for each
of the (a)(b)(c) and (d) categories:

category (a): $\dfrac{23!}{10!\,13!}$ = No. of hands with T1, T2 left

Category (b): $\dfrac{23!}{11!\,12!}$ = No. of hands with T1 left, T2 R8.

category (c) $\dfrac{23!}{11!\,12!}$ = No. of hands with T2 left, T1 R8.

Category (d) $\dfrac{23!}{12!\,11!}$ = No. of hands with T1, T2 Right

Part (a): P[T1, T2 are to the left] =

$$\dfrac{\dfrac{23!}{10!\,13!}}{\dfrac{23!}{10!\,13!}+\dfrac{23!}{11!\,12!}+\dfrac{23!}{11!\,12!}+\dfrac{23!}{12!\,11!}} = \dfrac{11}{50}$$

Part (b): P[T1, T2 are to the right] =

$$\dfrac{23!/11!\,12!}{\text{Sum as in (a)}} = \dfrac{13}{50}$$

Part (c): P[T1 and T2 are split] =

$$\left(23!/11!\,12! + 23!/12!\,11!\right)/\text{sum as in (a)} = 26/50$$

5

1.8 Take advantage of calculations in Table 1.1.

No. on first roll	Prob. of "Don't Pass" via particular no. on first roll
2	$1/36$
3	$2/36$
4	$3/36 \cdot 2/3 = 2/36$
5	$4/36 \cdot 3/5 = 1/15$
6	$5/36 \cdot 6/11 = 5/66$
7	0
8	$5/36 \cdot 6/11 = 5/66$
9	$4/36 \cdot 3/5 = 1/15$
10	$3/36 \cdot 2/3 = 2/36$
11	0
12	$1/36 \cdot 1/2 = 1/72$ (A stand off is like half win - half lose)

Ans. for part (a) $\Longrightarrow$ Total $= \dfrac{217}{440} \approx .49318$

Part (b). Ave. casino "take"

$$= 1 - 2 \cdot \frac{217}{440} = \frac{6}{440} \approx 1.36\%$$

1.9 Use total itemization approach. Let "0" denote boy, "1" denote girl. The 16 possibilities are:

```
0000     0100     1000   → 1100
0001   → 0101   → 1001     1101
0010   → 0110   → 1010     1110
→ 0011   0111     1011     1111
```

The 6 "favorable" events are denoted "→".
∴ P[2 boys, 2 girls] $= 6/16 = 3/8$

1.10 Use heuristic approach.

(a) Example of "typical" 3 zeros and 3 ones:

$$001101$$

$$P[\text{above combination}] = \left(\tfrac{1}{2}\right)^6$$

But there are $\binom{6}{3} = \dfrac{6!}{3!\,3!} = 20$ possible arrangements of 3 zeros and 3 ones.

$$\therefore\ P[\text{Exactly 3 zeros, 3 ones}] = 20 \cdot \left(\tfrac{1}{2}\right)^6 = \tfrac{5}{16}$$

(b) Using arguments similar to those in (a):

$$P[\text{Exactly 4 zeros, 2 ones}] = \binom{6}{4} \cdot \left(\tfrac{1}{2}\right)^6 = \tfrac{15}{64}$$

(c) Similarily,

$$P[\text{Exactly 5 zeros, 1 one}] = \binom{6}{5} \cdot \left(\tfrac{1}{2}\right)^6 = \tfrac{3}{32}$$

(d) $P[\text{6 zeros}] = \left(\tfrac{1}{2}\right)^6 = \tfrac{1}{64}$

1.11 Use heuristic approach just as in Prob 1.10.

Typical sequence with say 2 errors:

Error locations

$$[1\ 1\ \overset{\downarrow}{0}\ 1\ 0\ \overset{\downarrow}{0}\ 0\ 1\ 1]$$

n bits

$$P[\text{above situation}] = p^2 (1-p)^{n-2} \cdot \underbrace{}_{\substack{(\text{No. of arrangements} \\ \text{of 2 errors})}}$$

Generalization for k errors:

$$P[k \text{ errors}] = \binom{n}{k} p^k (1-p)^{n-k}$$

1.12

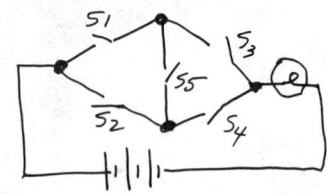

Let: "0" denote sw. open
"1" denote sw. closed

Then use a direct
itemization approach.

Position of Switches   Bulb "ON" or "OFF"

S5	S4	S3	S2	S1	
0	0	0	0	0	OFF
0	0	0	0	1	OFF
0	0	0	1	0	OFF
0	0	0	1	1	OFF
0	0	1	0	0	OFF
0	0	1	0	1	ON
0	0	1	1	0	OFF
0	0	1	1	1	ON
0	1	0	0	0	OFF
0	1	0	0	1	OFF
0	1	0	1	0	ON
0	1	0	1	1	ON
0	1	1	0	0	OFF
0	1	1	0	1	ON
0	1	1	1	0	ON
0	1	1	1	1	ON

} 7 "ON" CONDITIONS

REPEAT
FOR 16 MORE
POSITIONS

} 9 "ON" CONDITIONS

(a) $P[\text{BULB IS "ON"}] = \dfrac{7+9}{32} = \dfrac{1}{2}$

(b) $P[S5 \text{ CLOSED} | \text{BULB IS "ON"}]$

$= \dfrac{P[S5 \text{ CLOSED AND BULB "ON"}]}{P[\text{BULB "ON"}]} = \dfrac{9/32}{1/2} = \dfrac{9}{16}$

8

1.13

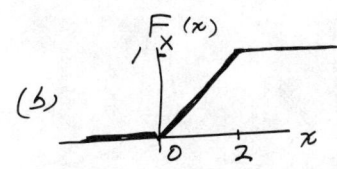

(a)

(b)

(c) $E(X) = \int_{-\infty}^{\infty} x f_X(x) dx = \int_0^2 x \cdot \frac{1}{2} dx = 1$

$E(X^2) = \int_{-\infty}^{\infty} x^2 f_X(x) dx = \int_0^2 x^2 \cdot \frac{1}{2} dx = 4/3$

$\text{Var } X = E(X^2) - [E(X)]^2 = 4/3 - 1 = 1/3$

1.14

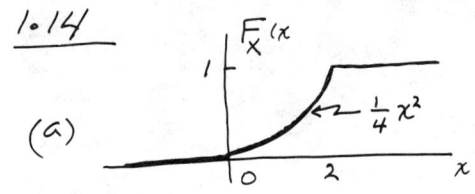

(a)

(b) $\text{Var } X = E(X^2) - [E(X)]^2$

$= \int_0^2 x^2 \cdot \frac{1}{2} x \, dx - \left[\int_0^2 x \cdot \frac{1}{2} x \, dx \right]^2$

$= 2 - (4/3)^2 = 2/9$

1.15 (a) Prob. density fcn. after sorting:

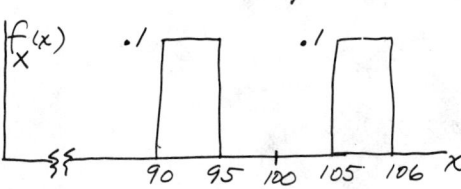

(b) Prob. that X is within 8 % is area between 92 and 95, and area between 105 and 108.

∴ $P[X \text{ is } 8\%] = .6$

1.16

$\sigma = 5$

90 100 110

Area between mean and 2σ is (from Table 19)

$.97725 - .5$

∴ Area for $\pm 2\sigma$
$$= 2(.97725 - .5)$$
$$= .9545$$

Finally, $P[\text{Sample is within} \pm 2\sigma] = .9545$

1.17

Let X = Time to failure. (exponentially distributed according to $\alpha e^{-\alpha x}$)

Then, $P[\text{Failure occurs between 0 and time } T]$
$$= \int_0^T \alpha e^{-\alpha x} dx = 1 - e^{-\alpha T}$$

(a) Average lifetime = $E(X)$
$$= \int_0^\infty x \cdot \alpha e^{-\alpha x} dx = \alpha \cdot \frac{1}{\alpha^2} = \frac{1}{\alpha}$$

For 10,000 hr lifetime, $\alpha = \frac{1}{10,000} \, hr^{-1}$

(b) $1 - e^{-\alpha T} = .01$

Solve for T:
$$e^{\alpha T} = 1/.99$$
$$T = \frac{1}{\alpha} \cdot \ln\left(\frac{1}{.99}\right) = 10,000 \cdot \ln\left(\frac{1}{.99}\right)$$
$$= 100.5 \, hrs.$$

1.18

(a) One coin

↑½   ↑½

───────────
  -1 0 +1

(b) 2 Coins

Events	Sum	Prob.
+1, +1	2	¼
+1, -1	0	} ½
-1, +1	0	
-1, -1	-2	¼

↑¼   ↑½   ↑¼
───────────────
  -2   0   +2

(c) 5 Coins

Events (Let "0" denote -1)	Sum	Prob.
0 0 0 0 0	-5	1/32

5 combinations with 1 zero
$\{$
0 0 0 0 1
0 0 0 1 0
⋮
1 0 0 0 0
$\}$ -3   5/32

$\binom{5}{2} = 10$ comb. with 2 zeros
$\{$
0 0 0 1 1
0 0 1 0 1
⋮
$\}$ -1   10/32

ETC.

	1	10/32
	3	5/32
	5	1/32

Final Result for 3 coins:

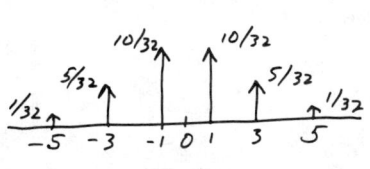

Part (d). Result for 10 coins:

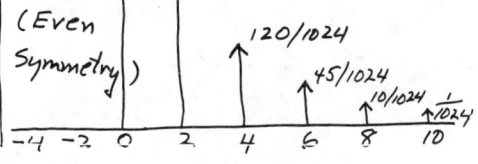

(Even Symmetry)

11

1.19   Initial Situation: <u>Player A</u>   <u>Player B</u>
                                2 coins   1 coin

Say A wins if coins match, B wins if not.

We are looking for the discrete prob. distribution
    that game ends in n trials

First, let $n = 0$. Obviously,
    $P[\text{Trials} = 0] = 0$

Next, let $n = 1$. Game will end if coins
    match (i.e., A takes B's last coin). As
    "match" or "mismatch" are equally likely,
    $P[\text{Trials} = 1] = \frac{1}{2}$

Next, look at $n = 2$ case. This can occur
    only if B wins at first step, which
    results in the following situation:
        <u>Player A</u>   <u>Player B</u>
         1 coin   2 coins

Now, if player B wins 2nd toss, the
game ends. Probability of this is $\frac{1}{2}$,
given the above situation. Therefore,
    $P[\text{Trials} = 2] = \frac{1}{2} \cdot \frac{1}{2} = \frac{1}{4}$

This can be continued ad infinitum.
Final prob. mass distribution is:

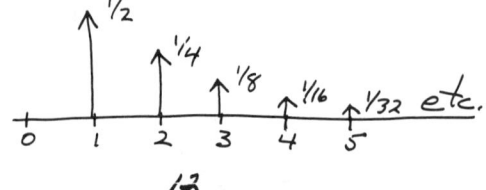

1.20

(a) We wish to find the probability that the sum is 7 or 11, given one die is "5". The other die must then be "2" or "6". The probability of this is clearly $\frac{1}{6} + \frac{1}{6} = \frac{1}{3}$

$$\therefore \quad P[\text{Instant. win} \mid \text{Red die is "5"}] = \frac{1}{3}$$

(b) The unconditional probability of obtaining 7 or 11 on 1st throw is $\frac{1}{6} + \frac{1}{18} = \frac{2}{9}$

1.21

Refer to Table 1.2 for probabilities of sums.

$$E(X) = 2 \cdot \frac{1}{36} + 3 \cdot \frac{2}{36} + 4 \cdot \frac{3}{36} + 5 \cdot \frac{4}{36} + 6 \cdot \frac{5}{36} + 7 \cdot \frac{6}{36}$$

$$+ 8 \cdot \frac{5}{36} + 9 \cdot \frac{4}{36} + 10 \cdot \frac{3}{36} + 11 \cdot \frac{2}{36} + 12 \cdot \frac{1}{36}$$

$$= \frac{252}{36} = 7$$

$$E(X^2) = (2)^2 \cdot \frac{1}{36} + (3)^2 \cdot \frac{2}{36} + \cdots + (12)^2 \cdot \frac{1}{36}$$

$$= \frac{1974}{36} = 54.833\cdots$$

$$\text{Var } X = E(X^2) - [E(X)]^2 = 54.833 - 7^2$$

$$= 5.8333\cdots$$

1.22

X \ Y	1	3	5	Marginal Prob. P(X)
1	1/18	1/18	1/18	1/6
3	1/18	1/18	1/6	5/18
5	1/18	1/6	1/3	5/9
Marginal Prob. P(Y)	1/6	5/18	5/9	

(a) Test for independence: Does $P(X,Y) = P(X) \cdot P(Y)$?

For example, try $X = 1$ and $Y = 1$.

$$P(X=1, Y=1) = 1/18 \quad \text{(from table)}$$

$$P(X=1) \cdot P(Y=1) = \frac{1}{6} \cdot \frac{1}{6} = \frac{1}{36}$$

We need not check any further. Independence criterion is **not** satisfied

(b) From the marginal probabilities, we have

$$P(Y=5) = 5/9$$

(c) $P(Y=5 \,|\, X=3) = \dfrac{P(Y=5, X=3)}{P(X=3)} = \dfrac{1/6}{5/18} = 3/5$

1.23   Refer to Fig. Problem 1.23 and first compute the joint probabilities:

$P(X=0, Y=0) = (.9)(.75) = .675$

$P(X=0, Y=1) = (.1)(.75) = .075$

$P(X=1, Y=0) = (.2)(.25) = .05$

$P(X=1, Y=1) = (.8)(.25) = .20$

1.23 (con't)

Part (c): Joint and marginal probabilities

X \ Y	0	1	
0	.675	.075	.75
1	.05	.20	.25
	.725	.275	

Part (b): Unconditional probabilities

$$P(Y=0) = .725$$
$$P(Y=1) = .275$$

Part (a): Conditional probabilities

$$P(X=0|Y=1) = \frac{.075}{.275} = \frac{3}{11}$$
$$P(X=0|Y=0) = \frac{.675}{.725} = \frac{27}{29}$$

1.24

Poisson distribution. Let $\alpha T = 2$. Then

$$P(k) = \frac{(2)^k}{k!} e^{-2}$$

Tabulated values:

k	P(k)
0	.135
1	.27
2	.27
3	.18
4	.09

etc.

1.24 (con't)

It is almost self-evident from the discussion that the mean of the distribution is λ. This can be justified formally as follows:

Write probability density function as

$$f(k) = \frac{\lambda^0}{!}e^{-\lambda}\delta(k) + \frac{\lambda'}{!}e^{-\lambda}\delta(k-1) + \frac{\lambda^2}{2!}e^{-\lambda}\delta(k-2) + \cdots$$

The Fourier transform of $f(k)$ is

$$\mathcal{F}[f(k)] = \frac{\lambda^0}{!}e^{-\lambda} + \frac{\lambda'}{!}e^{-\lambda}e^{-j\omega} + \frac{\lambda^2}{2!}e^{-\lambda}e^{-2j\omega} + \cdots$$

Now, replace $j\omega$ with $-j\omega$ and obtain Char. fcn.

$$\text{Char. fcn} = \psi(\omega) = e^{-\lambda} + \lambda e^{-\lambda}e^{j\omega} + \frac{\lambda^2}{2!}e^{-\lambda}e^{2j\omega} + \cdots$$

Next, form $d\psi/d\omega$ and evaluate at $\omega = 0$:

$$\frac{d\psi}{d\omega}\Bigg|_{\omega=0} = \left[0 + \lambda e^{-\lambda}e^{j\omega}(j) + \frac{\lambda^2}{2!}e^{-\lambda}e^{2j\omega}(2j) + \cdots \right]_{\omega=0}$$

$$= e^{-\lambda}\lambda j\left[1 + \frac{\lambda}{1!} + \frac{\lambda^2}{2!} + \cdots \right]$$

$$= e^{-\lambda}\lambda j\, e^{\lambda} = j\lambda$$

Now, from Sec. 1.8,

$$E(X) = \frac{1}{j}\frac{d\psi}{d\omega}\Bigg|_{\omega=0} = \frac{1}{j}\cdot j\lambda = \lambda$$

1.25 $P(k) = \dfrac{(\alpha T)^k}{k!} e^{-\alpha T}$

$\alpha = 3$ outages/yr , $T = 1$ yr, $\therefore \alpha T = 3$

$P[\text{Exactly zero outages}] = \dfrac{(3)^0}{1} e^{-3} \approx .0498$

1.26   Use Poisson dist. just as in Prob. 1.25.

(a) $\alpha = 1$ quake/decade , $T = 1$, $\therefore \alpha T = 1$

$P[\text{Exactly zero quakes}] = \dfrac{(1)^0}{1} e^{-1} \approx .368$

$P[\text{One or more quakes}] = 1 - .368 = .632$

(b) $\alpha = 1$ quake/dec., $T = 7$ dec. , $\alpha T = 7$

$P(0) = e^{-7} \approx .0009$

$P[\text{One or more quakes}] = 1 - .0009 = .9991$

1.27

Rayleigh density fcn $= \dfrac{r}{\sigma^2} e^{-r^2/2\sigma^2}$, $r > 0$

(a) Calculation of mean:

$$E(R) = \int_0^\infty r \cdot \dfrac{r}{\sigma^2} e^{-r^2/2\sigma^2} dr = \dfrac{1}{2} \dfrac{\sqrt{2\pi}\,\sigma}{\sigma^2} \int_{-\infty}^\infty \dfrac{r^2}{\sqrt{2\pi}\,\sigma} e^{-\frac{r^2}{2\sigma^2}} dr$$

The integral is just the variance of $N(0, \sigma^2)$. Therefore,

$$E(X) = \dfrac{\sqrt{2\pi}\,\sigma}{2\sigma^2} \cdot \sigma^2 = \sqrt{\dfrac{\pi}{2}}\,\sigma$$

Calculation of variance:

$$E(X^2) = \int_0^\infty \frac{r^3}{\sigma^2} e^{-r^2/2\sigma^2} dr$$

The above integral must be integrated by parts. The result is $2\sigma^2$.

$$\therefore \ Var X = E(X^2) - [E(X)]^2 = 2\sigma^2 - \left(\sqrt{\tfrac{\pi}{2}}\sigma\right)^2$$

$$= \sigma^2 \left(2 - \tfrac{\pi}{2}\right)$$

(b) The mode is the peak value of $f_R(r)$. Therefore, differentiate and set equal to zero.

$$\frac{df_R}{dr} = \frac{r}{\sigma^2} \cdot e^{-r^2/2\sigma^2} \cdot \left(-\tfrac{1}{\sigma^2}r\right) + \frac{1}{\sigma^2} \cdot e^{-r^2/2\sigma^2} = 0$$

Or $\quad -r^2 \frac{1}{\sigma^2} = -1$

Or $\quad r = \sigma \quad$ (Mode is σ)

<u>1.28</u>

Random variable X is number of heads
First, consider the realization X = 0.

Coin #1	Coin #2	Coin #3
Tail (.4)	Tail (.4)	Tail (.4)

$$\therefore \ P[X=0] = (.4)^3 = .064$$

1.28 (con't)

Next, consider $X=1$. Possible ways of success are:

$$\begin{array}{lll}
\text{Head (.6)} & \text{Tail (.4)} & \text{Tail (.4)} \\
\text{Tail (.4)} & \text{Head (.6)} & \text{Tail (.4)} \\
\text{Tail (.4)} & \text{Tail (.4)} & \text{Head (.6)}
\end{array}$$

$$\therefore\ P[X=1] = 3 \cdot (.4)^2 \cdot (.6) = .288$$

Similar analysis applies for $X=2$ and $X=3$.
Final result is:

X	Probability Mass Distribution
0	.064
1	.288
2	.432
3	.216

1.29

(a) $$P[R < R_0] = \int_0^{R_0} \frac{r}{\sigma^2} e^{-\frac{r^2}{2\sigma^2}} dr = 1 - e^{-R_0^2/2\sigma^2}$$

(b) To get CEP, let the result of part (a) equal $\frac{1}{2}$.

$$1 - e^{-R_0^2/2\sigma^2} = \frac{1}{2}$$

Now solve for R_0

$$e^{-R_0^2/2\sigma^2} = \frac{1}{2}$$

or $$R_0 = \sqrt{2\sigma^2 \ln 2} = \sigma \sqrt{2 \ln 2}$$

$$\approx 1.177\, \sigma$$

1.30

$f_Y(y) = \frac{1}{2}\delta(y) +$ "Right half" of $N(0, \sigma_x^2)$

Mean:

$$E(Y) = \int_{-\infty}^{\infty} y \cdot \frac{1}{2}\delta(y)\,dy + \int_0^{\infty} y \cdot \frac{1}{\sqrt{2\pi}\,\sigma_x}\,e^{-y^2/2\sigma_x^2}\,dy$$

$$= 0 + \frac{\sigma_x}{\sqrt{2\pi}}\int_0^{\infty} e^{-y^2/2\sigma_x^2}\left(\frac{y}{\sigma_x^2}dy\right) = \frac{\sigma_x}{\sqrt{2\pi}}$$

Variance:

$$E(Y^2) = \int_{-\infty}^{\infty} y^2 \cdot \frac{1}{2}\delta(y) + \int_0^{\infty} y^2 \cdot \frac{1}{\sqrt{2\pi}\,\sigma_x}\,e^{-y^2/2\sigma_x^2}\,dy$$

$$= 0 + \frac{1}{2}\sigma_x^2$$

$\therefore$ Var $Y = E(Y^2) - \{E(Y)\}^2 = \frac{\sigma_x^2}{2}\left[1 - \frac{1}{\pi}\right]$

1.31

$f_X = \begin{cases} e^{-x}, & x > 0 \\ 0, & x < 0 \end{cases}$

(a) $P(X \geq 2) = \int_2^{\infty} e^{-x}\,dx = e^{-2} \approx .135$

(b) $P(1 \leq X \leq 2) = \int_1^2 e^{-x} = e^{-1} - e^{-2} \approx .232$

(c) $E(X) = \int_0^{\infty} x e^{-x}\,dx = 1$

$E(X^2) = \int_0^{\infty} x^2 e^{-x}\,dx = 2$

Hint: These are just Laplace transform integrals, so tables can be used letting $s=1$ in the tables.

Var $X = E(X^2) - [E(X)]^2 = 2 - 1^2 = 1$

20

1.32 $f_{XY} = \begin{cases} .25 & , \text{ in square} \\ 0 & , \text{ otherwise} \end{cases}$

$f_X(x) = \begin{cases} \int_{-1}^{1} .25\, dy & , \text{ for } -1 < x < +1, \quad = .25y\big|_{-1}^{1} = \tfrac{1}{2} \\ 0 & , \text{ otherwise} \end{cases}$

Similarly, $f_Y(y) = \begin{cases} \tfrac{1}{2} & , \; -1 < y < 1 \\ 0 & , \text{ otherwise} \end{cases}$

∴ Test of $f_{XY} = f_X \cdot f_Y$ is satisfied and
X and Y are statistically independent.

1.33
$f_{XY} = \begin{cases} e^{-(x+y)} & , \text{ 1st quadrant} \\ 0 & , \text{ otherwise} \end{cases}$

Region (a)

(a) Integrate over region (a)

$P[X \le \tfrac{1}{2}] = \int_{0}^{\infty}\int_{0}^{1/2} e^{-x}\cdot e^{-y}\, dx\, dy = \int_{0}^{\infty} e^{-y}\int_{0}^{1/2} e^{-x}\, dx\, dy$

$= 1 - e^{-1/2} \approx .393$

(b) Integrate over region below x+y line.

$P[(X+Y) \le 1] = \int_{0}^{1}\int_{0}^{1-y} e^{-x}\cdot e^{-y}\, dx\, dy$

$x+y=1$

$= 1 - 2e^{-1} \approx .264$

(c) $P[(X \text{ or } Y) \ge 1] = 1 - \underset{\text{Area 1}}{\iint f_{XY}\, dx\, dy}$

$= 1 - (1-e^{-1})^2 \approx .60$

Area #2

Area #1

(d) $P[(X \text{ and } Y) \ge 1] = \underset{\text{Area 2}}{\iint f_{XY}\, dx\, dy}$

$= e^{-2} \approx .135$

1.34

In order to be stat. ind. $f_{XY} = f_X \cdot f_Y$

Check: First compute f_X:

$$f_X(x) = \int_{-\infty}^{\infty} f_{XY}(x,y)\, dy = \begin{cases} \int_0^{\infty} e^{-(x+y)}\, dy = e^{-x}, & \text{for } x \geq 0 \\ 0, & \text{for } x < 0 \end{cases}$$

Similarly,

$$f_Y(y) = \int_{-\infty}^{\infty} f_{XY}(x,y)\, dx = \begin{cases} e^{-y}, & \text{for } y \geq 0 \\ 0, & \text{for } y < 0 \end{cases}$$

Clearly, the product $f_X \cdot f_Y = f_{XY}$ for all x and y, so X and Y are statistically independent.

1.35

$$f_X(x) = \tfrac{1}{2} e^{-|x|} \quad , \quad f_Y = e^{-2|y|}$$

Define Z to be: $\quad Z = X + Y$

Then,

$$f_Z(3) = \int_{-\infty}^{\infty} f_X(u) \cdot f_Y(3-u)\, du$$

Rather than integrate above int. directly, use Fourier transform theory.

$$\mathcal{F}[f_Z] = \mathcal{F}[f_X] \cdot \mathcal{F}[f_Y]$$

$$= \frac{1}{\omega^2 + 1} \cdot \frac{4}{\omega^2 + 4}$$

Now use partial fraction expansion.

$$\mathcal{F}[f_Z] = \frac{4/3}{\omega^2 + 1} + \frac{-4/3}{\omega^2 + 4}$$

We now recognize the inverse of each term.

$$\therefore f_Z(3) = \frac{2}{3} e^{-|3|} - \frac{1}{3} e^{-2|3|}$$

1.36

The function $y = x^3 + 1$ is one-to-one, so solve for x in terms of y and use Eq. (1.14.6)

$$f_Y(y) = \left| \frac{dx}{dy} \right| f_X [h(y)]$$

Solving for $h(y)$: $x = \sqrt[3]{y-1} \triangleq h(y)$

(Note x must lie between -1 and 1, so $0 < y < 2$. Thus it is implied the $\sqrt[3]{}$ will taken so as to yield function shown darkened in the figure.)

For $0 < y < 2$, $f_X [h(y)] = \frac{1}{2}$

Also, $\left| \frac{dx}{dy} \right| = \left| \frac{1}{3}(y-1)^{-2/3} \right|$

$$\therefore f_Y(y) = \begin{cases} \frac{1}{6}(y-1)^{-2/3}, & 0 < y < 2 \\ 0, & \text{otherwise} \end{cases}$$

1.37

X and Y are zero mean, so U and V are zero mean. Let overscore denote expectation.

$$\overline{UV} = \overline{(2X+Y)(X-Y)} = 2\overline{X^2} + \overline{XY}^{\,0} - 2\overline{XY}^{\,0} - \overline{Y^2} = \sigma_x^2$$

$$\overline{U^2} = \overline{(2X+Y)^2} = 4\overline{X^2} + 4\overline{XY}^{\,0} + \overline{Y^2} = 5\sigma_x^2$$

$$\overline{V^2} = \overline{(X-Y)^2} = \overline{X^2} - 2\overline{XY}^{\,0} + \overline{Y^2} = 2\sigma_x^2$$

$$\rho = \frac{\overline{UV}}{\sqrt{\overline{U^2}} \cdot \sqrt{\overline{V^2}}} = \frac{\sigma_x^2}{\sigma_x \sqrt{5} \cdot \sigma_x \sqrt{2}} = \frac{1}{\sqrt{10}}$$

1.38

$$m_Y = \begin{bmatrix} 2 & 1 \\ 1 & -1 \end{bmatrix}\begin{bmatrix} 1 \\ 2 \end{bmatrix} + \begin{bmatrix} 1 \\ 1 \end{bmatrix} = \begin{bmatrix} 4 \\ -1 \end{bmatrix} + \begin{bmatrix} 1 \\ 1 \end{bmatrix} = \begin{bmatrix} 5 \\ 0 \end{bmatrix}$$

$$C_Y = \begin{bmatrix} 2 & 1 \\ 1 & -1 \end{bmatrix}\begin{bmatrix} 4 & 1 \\ 1 & 1 \end{bmatrix}\begin{bmatrix} 2 & 1 \\ 1 & -1 \end{bmatrix} = \begin{bmatrix} 21 & 6 \\ 6 & 3 \end{bmatrix}$$

1.39 Begin with the density functions

$$f_{\underline{x}}(x_1, x_2) = \frac{1}{2\pi \, |C_x|^{1/2}} \, e^{-\frac{1}{2}[(\underline{x}-\underline{m}_x)^T C_x^{-1}(\underline{x}-\underline{m}_x)]}$$

$$f_{X_1}(x_1) = \frac{1}{\sqrt{2\pi} \, \sigma_{x_1}} \, e^{-\frac{1}{2\sigma_{x_1}^2}(x_1 - m_{x_1})^2}$$

Then, from the definition of conditional density

$$f_{X_2|X_1} = \frac{f_{\underline{x}}(x_1, x_2)}{f_{X_1}(x_1)} = (\text{Const.}) \frac{e^{-\frac{1}{2}[(\underline{x}-\underline{m}_x)^T C_x^{-1}(\underline{x}-\underline{m}_x)]}}{e^{-\frac{1}{2\sigma_{x_2}^2}(x_1 - m_{x_1})^2}}$$

$$= (\text{Const.}) \frac{e^{-\frac{1}{2}[ax_1^2 + bx_1 x_2 + cx_2^2 + dx_1 + ex_2 + f]}}{e^{-\frac{1}{2}[a'x_1^2 + b'x_1 + c']}}$$

Clearly, the exponential term has the form
of a general quadratic in x_1 and x_2. To
show the Gaussian form, all we need to
do is to show that the quadratic form above
is the equivalent of the matrix form:

$$\begin{bmatrix} x_1 - m_1 \\ x_2 - m_2 \end{bmatrix}^T \begin{bmatrix} h_{11} & h_{12} \\ h_{12} & h_{22} \end{bmatrix}\begin{bmatrix} x_1 - m_1 \\ x_2 - m_2 \end{bmatrix} + \text{const.}$$

This is done by expanding the above and solving for
h_{11}, h_{12}, h_{22}, and the constant. This is routine from here on.

1.40

$$f_{XY}(x,y) = \begin{cases} 1, & \text{in triangle} \\ 0, & \text{otherwise} \end{cases}$$

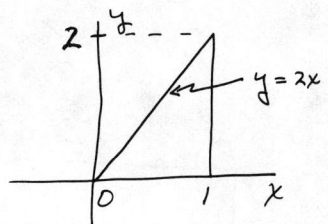

We must first find $f_Y(y)$:

$$f_Y(y) = \int_{-\infty}^{\infty} f_{XY}(x,y)\,dx = \int_{y/2}^{1} 1 \cdot dx = \begin{cases} 1 - y/2, & 0 < y < 2 \\ 0, & \text{otherwise} \end{cases}$$

Then

$$f_{X|Y} = \frac{1}{1 - y/2}, \quad 0 < y < 2 \text{ and } \frac{y}{2} < x < 1$$

In particular, where $y = .5$

$$f_{X|Y} = \frac{1}{1 - 1/4} = 4/3, \quad \frac{1}{4} < x < 1$$

The conditional mean is then

$$E[X \mid Y = .5]$$

$$= \int_{-1/4}^{1} x \cdot 4/3 \, dx = 5/8$$

CHAPTER 2

2.1

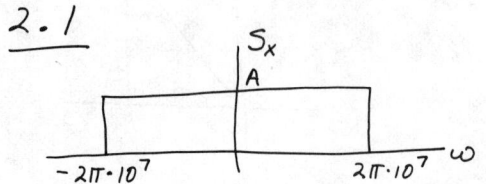

S_x
A

$-2\pi \cdot 10^7$ $2\pi \cdot 10^7$ ω

(a)
$$\frac{1}{2\pi}[Area] = Mean\ square\ value$$
$$\frac{1}{2\pi}[A \cdot 4\pi \cdot 10^7] = (100 \times 10^{-6}\ v)^2$$
$$\therefore\quad A = .5 \times 10^{-15}\ v^2/rad/sec$$

(b) Autocorrelation fcn. $= \mathcal{F}^{-1}[S_x(\omega)] = $ Sinc fcn.

10^{-8} $R_x(\tau)$

.05 μsec

τ

.1 μsec

$$R_x(\tau) = 10^{-8}\ \frac{\sin 2\pi \cdot 10^7 \tau}{2\pi \cdot 10^7 \tau}$$

2.2

Think of sliding $X(t)$ an amount τ and then multiplying the result times the original $X(t)$. Then average. The result is

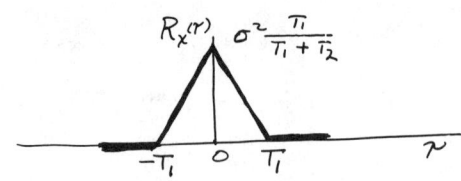

$R_x(\tau)$ $\sigma^2 \frac{T_1}{T_1 + T_2}$

$-T_1$ 0 T_1 τ

26

2.3

(a)

$$R(0) = \frac{1}{2}\left[\frac{1}{2} 1 \cdot 1 + \frac{1}{2} 0 \cdot 0\right] = \frac{1}{4}$$

Note that after a shift of an odd number of units, the pulses align with the "dead zones." Therefore, for $\tau = 1, 3, 5, 7$, etc., $R = 0$.

For a shift of 2, 4, 6, etc., there can be a nontrivial overlap of pulses. The following can occur:

$$0 \cdot 0 \qquad \text{Prob.} = \frac{1}{4}$$
$$0 \cdot 1 \qquad \text{Prob.} = \frac{1}{4}$$
$$1 \cdot 0 \qquad \text{Prob.} = \frac{1}{4}$$
$$1 \cdot 1 \qquad \text{Prob.} = \frac{1}{4}$$

Therefore, for $\tau = 2, 4, 6, 8, \cdots$ etc., $R = \frac{1}{2} \cdot \frac{1}{4} = \frac{1}{8}$

This leads to the following autocorrelation fcn.:

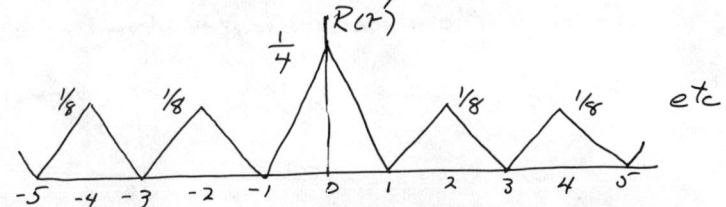

(b) Power spectral density function:

Resolve R into sum of 2 functions shown below.

$$S(j\omega) = \mathcal{F}\left[R(\tau)\right] = \mathcal{F}\left[\begin{array}{c}\text{Periodic}\\\text{Part}\end{array}\right] + \mathcal{F}\left[\begin{array}{c}\text{Triangular}\\\text{Part}\end{array}\right]$$

$$= \left[\frac{\pi}{8}\delta(\omega) + \frac{1}{2\pi}\delta(\omega+\omega_0) + \frac{1}{2\pi}\delta(\omega-\omega_0) + \frac{1}{2\pi \cdot 3^2}\delta(\omega+3\omega_0)\right.$$

$$\left. + \frac{1}{2\pi \cdot 3^2}\delta(\omega-3\omega_0) + \cdots \text{etc.}\right] + \frac{1}{8}\left(\frac{\sin\frac{\omega}{2}}{\frac{\omega}{2}}\right)^2$$

27

2.4

$$S(j\omega) = \delta(\omega) + \frac{1}{2}\delta(\omega - \omega_0) + \frac{1}{2}\delta(\omega + \omega_0) + e^{-2|\omega|}$$

$$R(\tau) = \mathcal{F}^{-1}[S(j\omega)]$$

$$= \frac{1}{2\pi} + \frac{1}{4\pi}e^{j\omega_0\tau} + \frac{1}{4\pi}e^{-j\omega_0\tau} + \frac{1}{2\pi} \cdot \frac{4}{\tau^2 + 2^2}$$

$$= \frac{1}{2\pi} + \frac{1}{2\pi}\cos\omega_0\tau + \frac{2}{\pi} \cdot \frac{1}{\tau^2 + 4}$$

2.5

Typical signal:

+4 units

−4 units

$$R = 4e^{-|\tau|}, \quad \text{thus } \sigma^2 = 4 \quad \text{or} \quad \sigma = 2$$

4 Units is 2× Standard Deviation
we find the areas indicated from Table 1.9.

shaded area
$$= 2(1 - .97725)$$

Normal Density

−2   0   1   2

$$\therefore P\{X \text{ exceeds } 4 \text{ units in magnitude}\} = .0455$$

2.6

(a) Typical waveform:

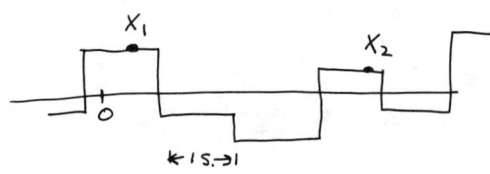

X_1 X_2

0

⊢ 1S. →|

First, let τ be large as shown (i.e., > than 1).
The unconditional densities:

$$f_{X_1}(x_1) = \frac{1}{\sqrt{2\pi}\,\sigma}e^{-\frac{1}{2\sigma^2}x_1^2}$$

$$f_{X_2}(x_2) = \frac{1}{\sqrt{2\pi}\,\sigma}e^{-\frac{1}{2\sigma^2}x_2^2}$$

2.6 (cont.)

If $|\gamma| > 1$, X_1 and X_2 are uncorrelated and $f_{X_1X_2} = f_{X_1} \cdot f_{X_2}$ when $|\gamma| < 1$, use the relative frequency concept of probability. Part of the time X_1 and X_2 are perfectly correlated; the other part of the time they are independent. Therefore, the joint density must be a blend of these two situations. The blending factor must be linear in γ for $|\gamma| < 1$. This leads to the following result (found by trial-and-error) for positive γ:

$$f_{X_1X_2}(x_1,x_2) = \begin{cases} \gamma \cdot \dfrac{1}{2\pi\sigma^2} e^{-\frac{1}{2\sigma^2}(x_1^2 + x_2^2)} + (1-\gamma)\cdot \dfrac{1}{\sqrt{2\pi}\sigma}\, \delta(x_1 - x_2) e^{-\frac{1}{4\sigma^2}(x_1^2 + x_2^2)} & 0 < \gamma < 1 \\[2mm] \dfrac{1}{2\pi\sigma^2} e^{-\frac{1}{2\sigma^2}(x_1^2 + x_2^2)} & |\gamma| > 1 \end{cases}$$

It can be verified that integrating $f_{X_1X_2}$ w.r.t. x_1 leads to the correct f_{X_2}, and vice versa.

(b) The above is not the standard bivariate form, so the process is *not* Gaussian.

2.7

$$E(X_1 X_2) = \int_{-\infty}^{\infty}\int_{-\infty}^{\infty} x_1 x_2 \left[\gamma \cdot \frac{1}{2\pi\sigma^2} e^{-\frac{1}{2\sigma^2}(x_1^2 + x_2^2)} \right. $$
$$\left. + (1-\gamma)\frac{1}{\sqrt{2\pi}\sigma}\, \delta(x_1 - x_2) e^{-\frac{1}{4\sigma^2}(x_1^2 + x_2^2)} \right] dx_1 dx_2$$

for $0 < \gamma < 1$

$$= (1-\gamma)\frac{1}{\sqrt{2\pi}\sigma} \int_{-\infty}^{\infty} x_2^2 e^{-\frac{1}{2\sigma^2}x_2^2} dx_2 = (1-\gamma)\sigma^2,$$

for $0 < \gamma < 1$

For $\gamma > 1$,

$$E(X_1 X_2) = 0 \ , \quad \therefore R(\gamma) \text{ is:}$$

2.8

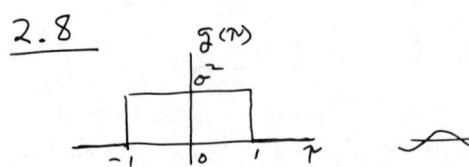 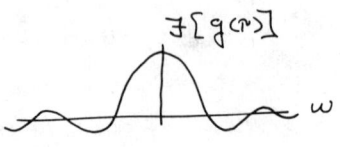

The Fourier transform of the above $g(\tau)$ is a sinc fcn.
This function has regions where it is negative.
But negative power spectral density is not possible
for real processes (see Eq. (2.7.9)). Therefore, the
rectangular "autocorrelation" shown is not possible.

2.9

$$X(t) = 2 \sin \omega t, \quad \omega \text{ is a R.V.}$$

Typical sample functions:

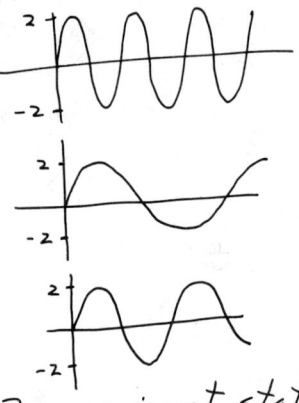

All have same
amplitude but
different frequencies
(determined by chance)

(a) Process is **not** stationary. Clearly, the "statistics" are
   different at $t=0$ than where t is positive.

(b) Process is **not** ergodic. (For example, the variance
   of "time samples" of any one function is not the
   same as the variance of "ensemble" samples at $t=0$.

(c) Process is deterministic in that any member of
   the ensemble is exactly predictable after 1 cycle.

<u>2.10</u>

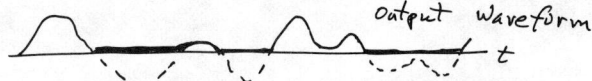

Output waveform

(a) Output is stationary. ("Statistics "do not change with time.)

(b) Output is nongaussian. Even the first-order density function is not normal (shown at right).

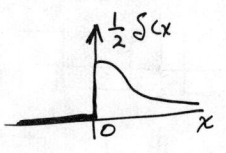

<u>2.11</u>

$X(t) = at + Y$; Y is $N(0, \sigma^2)$, a is known.

(a) Process is <u>not</u> stationary. This is obvious from the appearance of t explicitly in $X(t)$ expression.

(b) Process is <u>not</u> ergodic. Clearly, $X(t)$ at $t=0$ is $N(0, \sigma^2)$ and has zero mean. The average of time samples will depend on a and the sample times, so time and ensemble averaging are not the same.

<u>2.12</u>

$X(t) = at + Y$; Y is $N(0, \sigma^2)$, a is known.

Process is nonstationary so the a.c. fcn. will be a function of two parameters t_1 and t_2.

$$R_X(t_1, t_2) = E\left[(at_1 + Y)(at_2 + Y)\right]$$

$$= E\left[a^2 t_1 t_2 + at_1 Y + at_2 Y + Y^2\right]$$

$$= a^2 t_1 t_2 + \sigma^2$$

2.13

Typical ensemble of time
functions:

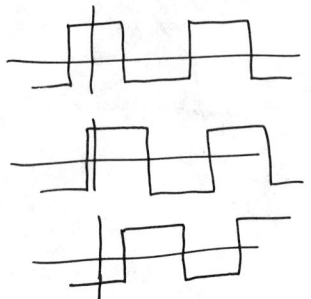

(a) The "origin" of the
square waves is random,
so the process is at
least <u>wide-sense</u> stationary
(see part (c).).

(b) The process is deterministic. (It is exactly
predictable after one cycle.)

(c) The autocorrelation function is obtained by
either time averaging $X(t)X(t+\tau)$ or ensemble
averaging $X(t_1) \cdot X(t_2)$, where t_1 and t_2 are
separated by τ. The result is:

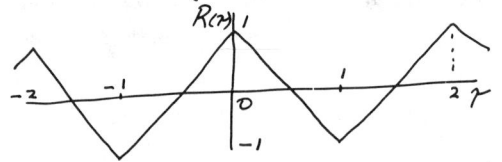

P.S.D. is obtained from the Four. transform of the
Four. series expansion of the $R(\tau)$ above.
The result:

2.14

$$R(\tau) = \sigma^2 e^{-\beta|\tau|} \cos \omega_0 \tau$$

(a) Refer to Appendix A.2 for Fourier transform.

$$\mathcal{F}[R(\tau)] = \frac{\sigma^2 \beta}{(\omega + \omega_0)^2 + \beta^2} + \frac{\sigma^2 \beta}{(\omega - \omega_0)^2 + \beta^2}$$

(b)

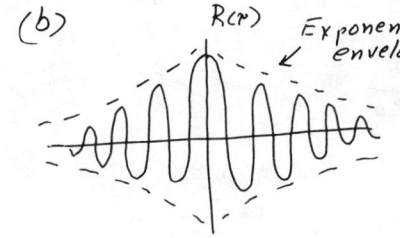

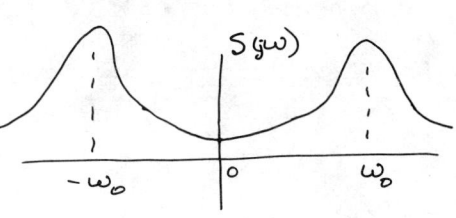

2.15

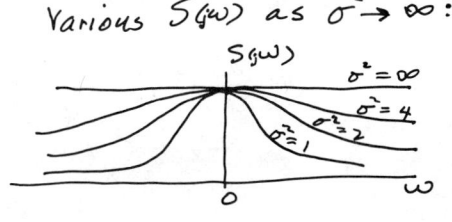

$$\text{Area} = 2 \int_0^\infty \sigma^2 e^{-\beta\tau} d\tau = \frac{2\sigma^2}{\beta}$$

Let $\dfrac{\sigma^2}{\beta} = \dfrac{1}{2}$ and then let $\sigma^2 \to \infty$

Area is then unity and invariant in the limiting process.

The power spectral density is:

$$S(j\omega) = \frac{2\sigma^2 \beta}{\omega^2 + \beta^2}$$

and where $\omega = 0$

$$S(0) = \frac{2\sigma^2}{\beta} = 1$$

Various $S(j\omega)$ as $\sigma^2 \to \infty$:

$$S(j\omega)$$

$\sigma^2 = \infty$

$\sigma^2 = 4$

$\sigma^2 = 2$

$\sigma^2 = 1$

2.16 $S_x = \dfrac{6\omega^2 + 12}{(\omega^2+4)(\omega^2+1)}$; now use a partial fraction expansion in ω^2:

$$S_x = \frac{4}{\omega^2+4} + \frac{2}{\omega^2+1}$$

S_x may now be integrated using standard integral tables to find mean square value.

$$E(X^2) = \frac{1}{2\pi}\int_{-\infty}^{\infty}\frac{4}{\omega^2+4}\,d\omega + \frac{1}{2\pi}\int_{-\infty}^{\infty}\frac{2}{\omega^2+1}\,d\omega = 1+1 = 2$$

2.17

$X(t)$ is stationary, so $Y(t)$ is also stationary.

(a) $R_y(\tau) = E[Y(t)\cdot Y(t+\tau)]$

$$= E[(aX(t)+b)(aX(t+\tau)+b)]$$
$$= E[a^2 X(t)X(t+\tau)] + E[ab\,X(t)]$$
$$\quad + E[ba\,X(t+\tau)] + E[b^2]$$
$$= a^2 R_x(\tau) + b^2 = a^2\sigma^2 e^{-\beta|\tau|} + b^2$$

(b) $R_{XY}(\tau) = E[X(t)\,Y(t+\tau)]$

$$= E[X(t)(aX(t+\tau)+b)]$$
$$= E[aX(t)X(t+\tau)] + E[X(t)\cdot b]$$
$$= a\,R_x(\tau) = a\sigma^2 e^{-\beta|\tau|}$$

34

2.18 Use the Fourier transform tables and the shifting theorem (in the time domain)

$$S_{xy}(j\omega) = \mathcal{F}[R_{xy}(\tau)] = \frac{1}{3}\left(\frac{\sin\frac{\omega}{2}}{\frac{\omega}{2}}\right)^2 e^{-j\frac{\omega}{2}}$$

2.19

(a) The rectification cuts off negative parts as shown at right.

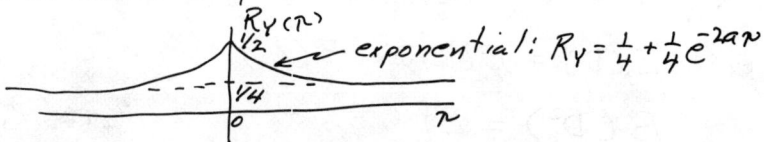

The output wave shape is the same as the input except the "lower level" is 0 rather than -1.

The mean square value can be seen to be ½. Therefore, $R_y(0) = \frac{1}{2}$

Also, the d-c value is ½. Therefore, $R_y(\infty) = \left(\frac{1}{2}\right)^2 = \frac{1}{4}$.

Auto correlation function is then:

exponential: $R_y = \frac{1}{4} + \frac{1}{4}e^{-2a\tau}$

(b) Crosscorrelation: $R_{xy}(\tau) = E[X(t)\,Y(t+\tau)]$

Now, the Y waveform is identical with the X waveform except reduced by a factor of ½ in amplitude and displaced up by ½.

$$\therefore \quad Y(t) = \frac{1}{2}X(t) + \frac{1}{2}$$

Using this in above equation for R_{xy}:

$$R_{xy}(\tau) = E\left[X(t)\cdot\left(\frac{1}{2}X(t) + \frac{1}{2}\right)\right]$$

$$= \frac{1}{2}R_x(\tau) = \frac{1}{2}e^{-2a|\tau|}$$

35

2.20

$$R_{xy}(\tau) = E\left[A \sin(\omega t + \theta) \cdot B \sin(\omega t + \omega \tau + \theta)\right]$$

But A and B are independent of θ. Therefore,

$$R_{xy}(\tau) = \underbrace{E[A \cdot B]}_{\text{Covar. } (A \cdot B)} \cdot \underbrace{E\left[\sin(\omega t + \theta) \cdot \sin(\omega t + \theta + \omega \tau)\right]}_{\frac{1}{2} \cos \omega \tau}$$

$$\therefore \; R_{xy}(\tau) = \rho \sigma^2 \cdot \frac{1}{2} \cos \omega \tau = \rho \frac{\sigma^2}{2} \cos \omega \tau$$

2.21

ℓ_i	Prob
ℓ	$\frac{1}{2}$
$-\ell$	$\frac{1}{2}$

Probability mass dist.

$$D = \ell_1 + \ell_2 + \ell_3 \cdots \ell_N$$

$$E(D) = E(\ell_1) + E(\ell_2) + \cdots = 0$$

$$E(D^2) = E\left[\ell_1^2 + \ell_2^2 + \cdots + 2\ell_1 \ell_2 + 2\ell_1 \ell_3 + \cdots \right]$$

$$= \ell^2 + \ell^2 + \cdots + 0 + 0 + \cdots$$

$$= N\ell^2$$

$$\therefore \; \text{Var } D = E(D^2) - [E(D)]^2 = N\ell^2$$

2.22

$$Z(t) = X(t) \, Y(t)$$

$$R_Z(\tau) = E\left[X(t) \, Y(t) \, X(t+\tau) \, Y(t+\tau)\right]$$

$$= E\left[X(t) X(t+\tau) \cdot Y(t) \, Y(t+\tau)\right]$$

Denote R.V.s $X(t)$ as X_1, $X(t+\tau)$ as X_2, $Y(t)$ as Y_1, $Y(t+\tau)$ as Y_2

Random Variables X_1 and X_2 are independent of Y_1 and Y_2. Therefore the joint density factors:

$$f_{X_1 X_2 Y_1 Y_2} = f_{X_1 X_2} \cdot f_{Y_1 Y_2}$$

Also,

$$E[X_1 X_2 Y_1 Y_2] = \iiiint x_1 x_2 \, y_1 y_2 \, f_{X_1 X_2}(x_1 x_2) f_{Y_1 Y_2}(y_1 y_2) \, dx_1 dx_2 dy_1 dy_2$$

$$= \iint x_1 x_2 f_{X_1 X_2}(x_1 x_2) dx_1 dx_2 \cdot \iint y_1 y_2 f_{Y_1 Y_2}(y_1 y_2) dy_1 dy_2 = \underbrace{E(X_1 X_2)}_{E(Y_1 Y_2)}$$

Returning now to the equation for R_z,

$$R_z(\tau) = E[X(t) X(t+\tau)] \cdot E[Y(t) \cdot Y(t+\tau)]$$
$$= R_x(\tau) \, R_Y(\tau)$$

Finally, use the complex convolution theorem and obtain

$$S_z(s) = \mathcal{F}\{R_z(\tau)\} = \frac{1}{2\pi j} \int_{-j\infty}^{j\infty} S_x(w) S_y(s-w) \, dw$$

$$S_x(j\omega) = \frac{1}{(1+\omega^2)^2}$$

If an elaborate set of Four. transform pairs were available, this would be a simple table-look-up problem. However, such tables are usually not readily available. Therefore, we will use a partial fraction expansion approach and, for variety, we will change to the "s" notation.

$$S_x(s) = \frac{1}{(-s^2+1)^2} , \quad \begin{array}{l}\text{second order roots}\\ \text{at } +1 \text{ and } -1\end{array}$$

2.23 (cont.)

Next, write S_x as a sum of positive- and negative-time parts, i.e.,

$$S_x = \frac{As+B}{(s+1)^2} + \frac{Cs+D}{(-s+1)^2} \quad , \quad A, B, C, D \text{ are to be determined.}$$

$\underbrace{\phantom{\frac{As+B}{(s+1)^2}}}$ Pos-time Part $\qquad$ $\underbrace{\phantom{\frac{Cs+D}{(-s+1)^2}}}$ Neg. time Part

Now rewrite S_x again as a partial fraction expansion.

$$S_x = \frac{C_1}{(s+1)^2} + \frac{C_2}{(s+1)} + \frac{C_3}{(-s+1)^2} + \frac{C_4}{(-s+1)}$$

$\underbrace{\phantom{\frac{C_1}{(s+1)^2}}}$ Pos.time part $\qquad$ $\underbrace{\phantom{\frac{C_3}{(-s+1)^2}}}$ Neg. Time part

Next, find C_1 and C_2 by usual methods

$$C_1 = \frac{1}{(s+1)^2(-s+1)^2} \cdot (s+1)^2 \Big|_{s=-1} = \frac{1}{4}$$

$$C_2 = \frac{d}{ds}\left[\frac{1 \cdot (s+1)}{(s+1)^2(-s+1)^2}\right]\Big|_{s=-1} = \frac{1}{4}$$

$$\therefore \quad R_x(\tau) = \frac{1}{4}|\tau| e^{-|\tau|} + \frac{1}{4} e^{-|\tau|}$$

2.24

X is zero-mean because $R_x(\tau) \to 0$ as $\tau \to \infty$.

X_1 and X_2 are separated by 1 unit.

Therefore, $E[X_1 X_2] = R_x(1) = 4e^{-1}$

The bivariate normal density is then

$$f_{X_1 X_2} = \frac{1}{2\pi|C|^{1/2}} e^{-\frac{1}{2}(x^T C^{-1} x)}$$

where

$$x = \begin{bmatrix} x_1 \\ x_2 \end{bmatrix} \quad , \quad C = \begin{bmatrix} 4 & 4e^{-1} \\ 4e^{-1} & 4 \end{bmatrix}$$

2.25 Typical pseudorandom noise (PN) code:

0 1 1 0 1   0 1 1

|← 1 sec |
1.023×10⁶

$\underbrace{\qquad\qquad}$ 1023 Bits (Repetition period = 1 ms.)

(a) Time autocorrelation function. By analogy to Example 2.13 we have

$R(\tau)$

$\frac{1 \times 10^{-6}}{1.023}$ sec

10^{-3} sec

$-1/1023$

(b) The Four. Transform of $R(\tau)$ give power spectral density. Note $R(\tau)$ is periodic, so $S(\omega)$ will consist of a sequence of impulse functions

S

(sinc fcn.)² envelope

Line spacing = 1 kH₃

-1 0 1 2 3

↑ Approx. 1 MH₃

kH₃

(Note that the frequency scale is in kH₃ for convenience and that the plot is of power spectral density, not "spectrum" to the first power.)

(c) RF Spectrum is similar to (b) except translated up to carrier frequency.

very closely spaced lines (only .001 MH₃ apart.)

1575.42 MH₃

2.26

Following Example 2.14, we have

$$\frac{\text{Std. Dev. of } V(\tau)}{\sigma^2} \leq \sqrt{\frac{2}{\beta T}}$$

In this case, critical value of T is obtained:

$$(.05) = \sqrt{\frac{2}{(.1) T}}$$

Or

$$T = \frac{2}{(.1)(.05)^2} = 8000 \text{ sec}$$

∴ Record length should be at least 8000 sec to achieve "5%" accuracy. (This is somewhat greater than 2 hours!)

2.27

Check on bandwidth assumption:

PSD is $\frac{2\sigma^2 \beta}{w^2 + \beta^2}$, $\beta = .1$ rad/sec; Half power point is .1 Rad/sec, or $\frac{.1}{2\pi}$ Hz.

(a) Assume sampling rate is $2W$, or $(2)(.1) = .2$ Hz
Sampling interval $= \frac{1}{.2} = 5$ sec
∴ Number of samples $= 8000/5 = 1600$

(b) We need a minimum of 1600 samples for a record length of 8000 sec. Check various powers of 2:

$$2^9 = 512, \quad 2^{10} = 1024, \quad 2^{11} = 2048$$

It appears that $N = 2048$ would be a good choice. (This is slightly higher sampling rate than needed, but no harm is done.)

(c) If the designer has a choice, he should increase the record length (by a factor of $2048/1600$ in this case.) Theoretically, over sampling produces no additional information -- increasing the record length does!!

2.28   We wish to show: (For zero-mean Gaussian case)

$$E(X_1 X_2 X_3 X_4) = E(X_1 X_2) E(X_3 X_4) + E(X_1 X_3) E(X_2 X_4) + E(X_1 X_4) E(X_2 X_3)$$

This problem involves a considerable amount of algebra, and there is no obvious way to avoid it.

Begin by noting that

$$E(X_1 X_2 X_3 X_4) = \frac{\partial^4 \psi}{\partial \omega_1 \partial \omega_2 \partial \omega_3 \partial \omega_4} \cdot (-j)^4 \Big|_{\omega_1 = \omega_2 = \omega_3 = \omega_4 = 0}$$

Next, write ψ in convenient matrix form and differentiate.

$$\frac{\partial \psi}{\partial \omega_1} = \frac{\partial}{\partial \omega_1} \left(e^{-\frac{1}{2} \omega^T C_x \omega} \right) = e^{-\frac{1}{2} \omega^T C_x \omega} \cdot \frac{\partial}{\partial \omega_1} \left(-\frac{1}{2} \omega^T C_x \omega \right)$$

Now, let $C_x = \begin{bmatrix} C_{11} & C_{12} & \cdots & C_{14} \\ C_{21} & C_{22} & \cdots \\ \cdots & \cdots & \cdots \end{bmatrix}$, write $-\frac{1}{2} \omega^T C_x \omega$ in component form, and differentiate w.r.t. ω_1. This then yields

$$\frac{\partial \psi}{\partial \omega_1} = - e^{-\frac{1}{2} \omega^T C_x \omega} (C_{11} \omega_1 + C_{12} \omega_2 + C_{13} \omega_3 + C_{14} \omega_4)$$

In a similar manner, form $\partial^2 \psi / \partial \omega_1 \partial \omega_2$. This works out to be:

$$\frac{\partial}{\partial \omega_2} \left(\frac{\partial \psi}{\partial \omega_1} \right) = - \left[e^{-\frac{1}{2} \omega^T C_x \omega} \right] \left[C_{12} + (C_{11} \omega_1 + C_{12} \omega_2 + C_{13} \omega_3 + C_{14} \omega_4) \cdot (-C_{21} \omega_1 + C_{22} \omega_2 + C_{23} \omega_3 + C_{24} \omega_4) \right]$$

Now, at this point anticipate that we will not differentiate w.r.t. ω_1 and ω_2 further, and that eventually ω_1 and $\omega_2 \to 0$. Therefore, terms that are linear in ω_1 and ω_2 can be dropped. Proceeding along this line of reasoning (and simplified algebra) leads eventually to:

$$\frac{\partial^4 \psi}{\partial \omega_1 \partial \omega_2 \partial \omega_3 \partial \omega_4} \Big|_{\substack{\omega_1 = 0 \\ \omega_2 = 0 \\ etc.}} = C_{12} C_{34} + C_{13} C_{24} + C_{14} C_{23}$$

$$= E(X_1 X_2) \cdot E(X_{34}) + \cdots etc.$$

2.29

(a) Refer to figure at right (N even)

From Eq 2.17.6:

$$\mathcal{G}_0 = \frac{1}{N} \sum_k g_k e^{-j \cdot 0} = \frac{1}{N} \sum_k g_k = \text{Real no. (for } g_k \text{ real)}$$

$$\mathcal{G}_{\frac{N}{2}} = \frac{1}{N} \sum_k g_k e^{-j\pi k} = \frac{1}{N} [g_0 - g_1 + g_2 - g_3 \cdots] = \text{Real no.}$$

However, it should be clear that $\mathcal{G}_1$ and $\mathcal{G}_{\frac{N}{2}-1}$ are, in general, complex. So are $\mathcal{G}_2$ and $\mathcal{G}_{\frac{N}{2}-2}$, etc. Therefore, the total number of real numbers required to describe $\mathcal{G}_0, \mathcal{G}_1, \mathcal{G}_2, \cdots \mathcal{G}_{N/2}$ is

$$\text{Number of real elements} = 1 + 2(\tfrac{N}{2} - 1) + 1 = N$$

(b) Refer to figure at right (N odd)

Just as in part (a)

$$\mathcal{G}_0 = \frac{1}{N} \sum_k g_k = \text{Real no.}$$

Clearly, though, $\mathcal{G}_1, \mathcal{G}_2 \cdots \mathcal{G}_{\frac{N}{2}-1/2}$ are in general complex. Therefore, the number of real elements required to describe $\mathcal{G}_m$ is

$$1 + \left(\frac{N}{2} - \frac{1}{2}\right) \cdot 2 = N$$

It can now be seen that only N real numbers are required to describe $\mathcal{G}_m$ for N either even or odd.

2.30   The figure shown is a direct implementation of Eq 2.12.1 :

$$S(t) = X(t) \cos \omega_c t - Y(t) \sin \omega_c t$$

where $X(t)$ and $Y(t)$ are stationary, independent baseband Gaussian processes.

(a)
$$R_s = E[S(t_1) \cdot S(t_2)]$$

$$= E[(X(t_1) \cos \omega_c t_1 - Y(t_1) \sin \omega_c t_1) \cdot$$
$$(X(t_2) \cos \omega_c t_2 - Y(t_2) \sin \omega_c t_2)]$$

$$= E[X(t_1)X(t_2) \cos \omega_c t_1 \cos \omega_c t_2 + Y(t_1)Y(t_2) \sin \omega_c t_1 \sin \omega_c t_2]$$

Now note that X and Y are stationary with identical autocorrelation functions, say R_X. Thus

$$R_s = R_x(t_2-t_1)[\cos \omega_c t_1 \cos \omega_c t_2 + \sin \omega_c t_1 \sin \omega_c t_2]$$

$$= R_x(t_2-t_1) \cos \omega_c(t_2-t_1)$$

Now replace t_2-t_1 with τ and obtain

$$R_s(\tau) = R_x(\tau) \cos \omega_c \tau$$

(b) Let sine channel be omitted. Then

$$S(t) = X(t) \cos \omega_c t$$

$$R_s = E[X(t_1) \cos \omega_c t_1 \cdot X(t_2) \cos \omega_c t_2]$$

$$= \{E[X(t_1)X(t_2)]\} \cdot \cos \omega_c t_1 \cos \omega_c t_2$$

$$= R_x(t_2-t_1) \cos \omega_c t_1 \cos \omega_c t_2$$

Note "cos·cos" term does _not_ reduce to a function of t_2-t_1, so process is not stationary.

CHAPTER 3

3.1

From Fourier transform theory

$$X(t) = \int_{-\infty}^{\infty} g(u) f(t-u) \, du$$

(In the above, we assume that $g(u)$ is Four. transformable and that the integral exists for given f.)

Similarly
$$X(t+\tau) = \int_{-\infty}^{\infty} g(v) f(t+\tau-v) \, dv$$

Therefore
$$E\{X(t)X(t+\tau)\} = \int_{-\infty}^{\infty}\int_{-\infty}^{\infty} g(u) \, g(v) \, E\{f(t-u) f(t+\tau-v)\} \, du \, dv$$

Or
$$R_x(\tau) = \int_{-\infty}^{\infty}\int_{-\infty}^{\infty} g(u) \, g(v) \, R_f(-u+v-\tau) \, du \, dv$$

Now, form the Four. transform of both sides of eq.

$$S_x(j\omega) = \int_{-\infty}^{\infty} \left[\int_{-\infty}^{\infty}\int_{-\infty}^{\infty} g(u) g(v) R_f(-u+v-\tau) \, du \, dv \right] e^{-j\omega\tau} \, d\tau$$

Next, interchange order of integration. This yields

$$S_x(j\omega) = \int_{-\infty}^{\infty}\int_{-\infty}^{\infty} g(u) g(v) S_f(j\omega) e^{j\omega u} e^{-j\omega v} \, du \, dv$$

$$= S_f(j\omega) \int_{-\infty}^{\infty} g(u) e^{j\omega u} \, du \int_{-\infty}^{\infty} g(v) e^{-j\omega v} \, dv$$

$$= S_f(j\omega) \, |G(j\omega)|^2$$

3.2 Use integral tables p.126 for all parts.

(a)
$$\bar{x^2} = \frac{1}{2\pi j} \int_{-j\infty}^{j\infty} A \cdot \frac{Ts}{(1+Ts)^2} \cdot \frac{T(-s)}{(1+T(-s))^2} \, ds \quad \text{(Overscore means "average")}$$

$$= \frac{1}{2\pi j} \int_{-j\infty}^{j\infty} \frac{\sqrt{A} Ts}{T^2 s^2 + 2Ts + 1} \cdot \left(\substack{\text{same} \\ \text{with} \\ s \to -s}\right) \cdot ds$$

Using the tables:

$$m = 2, \quad c_1 = \sqrt{A}T \qquad d_2 = T^2$$
$$\qquad\qquad c_0 = 0 \qquad\qquad d_1 = 2T$$
$$\qquad\qquad\qquad\qquad\qquad d_0 = 1$$

$$\therefore \bar{x^2} = \frac{c_1^2 d_0}{2 d_0 d_1 d_2} = \frac{AT^2}{2 \cdot 2T \cdot T^2} = \frac{A}{4T}$$

(b)
$$G(s) = \frac{\omega_0^2}{s^2 + 2\varsigma\omega_0 s + \omega_0^2}$$

Following the same procedure as in (a):

$$m = 2 \qquad c_1 = 0 \qquad\qquad d_2 = 1$$
$$\qquad\qquad c_0 = \sqrt{A}\,\omega_0^2 \qquad d_1 = 2\varsigma\omega_0$$
$$\qquad\qquad\qquad\qquad\qquad d_0 = \omega_0^2$$

$$\therefore \bar{x^2} = I_2 = \frac{c_0^2 d_2}{2 d_0 d_1 d_2} = \frac{A\omega_0^4}{2 \cdot \omega_0^2 \cdot 2\varsigma\omega_0} = \frac{A\omega_0}{4\varsigma}$$

(c)
$$G(s) = \frac{s+1}{(s+2)^2} = \frac{s+1}{s^2 + 4s + 4}$$

$$m = 2 \qquad c_1 = \sqrt{A} \qquad d_2 = 1$$
$$\qquad\qquad c_0 = \sqrt{A} \qquad d_1 = 4$$
$$\qquad\qquad\qquad\qquad\qquad d_0 = 4$$

$$\bar{x^2} = I_2 = \frac{(d_0 + d_2)A}{2 d_0 d_1 d_2} = \frac{(4+1)A}{2 \cdot 4 \cdot 4 \cdot 1} = \frac{5}{32}A$$

3.3

First, find voltage V' in terms of V_1.

Use voltage divider rule:

$$V' = \frac{\frac{(1\times10^6)(1/10^{-6}s)}{1\times10^6 + 1/10^{-6}s}}{1\times10^6 + \frac{(1\times10^6)(1/10^{-6}s)}{1\times10^6 + 1/10^{-6}s}} \, V_1 = \frac{1}{s+2} \, V_1$$

Obviously, $V_2 = \frac{1}{2} V'$

$\therefore$ Transfer function $G(s) = \dfrac{V_2(s)}{V_1(s)} = \dfrac{\frac{1}{2}}{s+2}$

Mean square value:

$$\overline{V_2^2} = \frac{1}{2\pi j} \int_{-j\infty}^{j\infty} A \cdot \frac{1/2}{s+2} \cdot \frac{1/2}{-s+2} \, ds$$

(Over score means average.)

Using integral tables, p.126, leads to

$$n=1, \qquad c_0 = \sqrt{A}/2, \qquad d_1 = 1$$
$$d_0 = 2$$

$$\therefore \overline{V_2^2} = I_1 = \frac{c_0^2}{2d_0d_1} = \frac{A/4}{2\cdot2\cdot1} = \frac{A}{16}$$

3.4

First, find the transfer function for feedback system.

$$G(s) = \frac{X(s)}{F(s)} = \frac{\frac{1}{s}}{1 + \frac{1}{s}\cdot K} = \frac{1}{s+K}$$

3.4 (cont.)

(a) The spectral density function is then

$$P_x'(s) = \frac{2\sigma^2\beta}{-s^2+\beta^2} \cdot \frac{1}{s+k} \cdot \frac{1}{-s+k}$$

Or, after spectral factorization

$$S_x(s) = \frac{\sqrt{2\sigma^2\beta}}{(s+\beta)(s+k)} \cdot \frac{\sqrt{2\sigma^2\beta}}{(-s+\beta)(-s+k)}$$

(b) Mean square value is obtained from tables, p. 126.

$$n = 2 \qquad c_1 = 0 \qquad\qquad d_2 = 1$$
$$\qquad\qquad c_0 = \sqrt{2\sigma^2\beta} \qquad d_1 = \beta + k$$
$$\qquad\qquad\qquad\qquad\qquad d_0 = \beta k$$

$$E[x^2] = \frac{c_0^2 d_2}{2 d_0 d_1 d_2} = \frac{2\sigma^2\beta}{2\cdot\beta k\cdot(\beta+k)} = \frac{\sigma^2}{k(\beta+k)}$$

3.5

(a) Spectral density (in terms of s):

$$S_x(s) = \frac{2\sigma^2\beta}{-s^2+\beta^2} \cdot \frac{1-T_1 s}{1+T_2 s} \cdot \frac{1+T_1 s}{1-T_2 s}$$

(b) To get mean square value, factor S_x, being careful to group left-half plane poles and $\underline{zeros}$ together.

$$S_x(s) = \frac{\sqrt{2\sigma^2\beta}\,(T_1 s+1)}{(s+\beta)(T_2 s+1)} \cdot \begin{bmatrix} \text{Mirror image} \\ \text{with right-half} \\ \text{poles and zeros} \end{bmatrix}$$

Now use the tables on p. 126.

$$n = 2 \qquad c_1 = T_1\sqrt{2\sigma^2\beta} \qquad d_2 = T_2$$
$$\qquad\qquad c_0 = \sqrt{2\sigma^2\beta} \qquad d_1 = 1+\beta T_2$$
$$\qquad\qquad\qquad\qquad\qquad d_0 = \beta$$

$$E(x^2) = \frac{T_1^2(2\sigma^2\beta)\beta + 2\sigma^2\beta T_2}{2\cdot\beta\cdot(1+\beta T_2)\cdot T_2} = \sigma^2\frac{(T_2+T_1^2\beta)}{(T_2+T_2^2\beta)}$$

47

3.6

This problem is more easily worked with real frequency ω than in the s domain.

First, use Fourier transform tables to get $S_f(j\omega)$. Input spectral function is then

$$S_f(j\omega) = \frac{\sigma^2}{\beta}\left[\frac{\sin\left(\frac{\omega}{2\beta}\right)}{\left(\frac{\omega}{2\beta}\right)}\right]^2$$

Output spectral function is then $S_f(j\omega)\,G(j\omega)\,G(-j\omega)$, or

$$S_x(j\omega) = \frac{\sigma^2}{\beta}\left[\frac{\sin\left(\frac{\omega}{2\beta}\right)}{\left(\frac{\omega}{2\beta}\right)}\right]^2 \frac{1}{1+T^2\omega^2}$$

This must be integrated to get mean square value. One way to make the problem workable with conventional integral tables is to rewrite S_x as

$$S_x = \frac{4\sigma^2\beta}{T^2}\left[\sin\left(\frac{\omega}{2\beta}\right)\right]^2\left[\frac{1}{\omega^2\left(\omega^2+\frac{1}{T^2}\right)}\right]$$

Now use partial fraction expansion on last term:

$$\frac{1}{\omega^2\left(\omega^2+\frac{1}{T^2}\right)} = \frac{T^2}{\omega^2} + \frac{-T^2}{\omega^2+\frac{1}{T^2}}$$

Mean square value may now be written as

$$E(x^2) = \frac{1}{2\pi}\cdot\frac{4\sigma^2\beta}{T^2}\cdot T^2\int_{-\infty}^{\infty}\left[\frac{\sin^2\left(\frac{\omega}{2\beta}\right)}{\omega^2} - \frac{\sin^2\left(\frac{\omega}{2\beta}\right)}{\omega^2+\frac{1}{T^2}}\right]d\omega$$

First term will be found in integral tables such as Chemical Rubber Co. Handbook of Chemistry and Physics. Second term can be written as a sum of 2 terms by writing $\sin^2\left(\frac{\omega}{2\beta}\right)$ as $\frac{1}{2}-\frac{1}{2}\cos\left(\frac{\omega}{\beta}\right)$. Each term can now be integrated. The final result is:

$$E(x^2) = \sigma^2\left[1 - T\beta\left(1 - e^{-\frac{1}{\beta T}}\right)\right]$$

3.7

$$S_f = 2RkT$$

(a) Evaluate S_f first:
$$S_f = (2)(10^6)(1.38 \times 10^{-23})(290) = 800.4 \times 10^{-17} \text{ volt}^2/\frac{rad}{sec}$$

Transfer function:
$$G(s) = \frac{\frac{1}{cs}}{R + \frac{1}{cs}} = \frac{1}{Ts+1}$$

Output Spectral function:
$$S_x(s) = \frac{800.4 \times 10^{-17}}{(Ts+1)(-Ts+1)}$$

Mean square value is found using integral tables, p.126.
$$E(x^2) = \frac{1}{2\pi j}\int_{-j\infty}^{j\infty} S_x(s)ds \approx 400 \times 10^{-11} \text{ v}^2$$

Or, rms value is:
$$\text{RMS Output voltage} \approx \sqrt{400 \times 10^{-11}} \approx 63.2 \text{ } \mu volt$$

(b) Half-power frequency $= \frac{1}{RC} = \frac{1}{10^6 \cdot 10^{-12}} = 10^6 \text{ rad/sec}$
$$\approx 159 \text{ kHz}$$

(c) The cutoff frequency is many orders of magnitude below infrared, so the flat approximation is quite good.

3.8

Note this filter reaches steady-state in time T.
Thus

$$E(x^2) = \int_0^T \int_0^T g(u) g(v) \, A\, \delta(u-v) \, du\, dv$$

$$= \int_0^T \int_0^T 1 \cdot 1 \cdot A\, \delta(u-v) \, du\, dv = AT$$

The same result can be obtained using spectral methods.

$$G(j\omega) = T \; \frac{\sin\left(\frac{\omega T}{2}\right)}{\left(\frac{\omega T}{2}\right)} \, e^{-j\frac{\omega T}{2}}$$

or

$$|G(j\omega)|^2 = T^2 \left[\frac{\sin\left(\frac{\omega T}{2}\right)}{\left(\frac{\omega T}{2}\right)}\right]^2$$

Output spectral function can now be integrated
with the result: $E(x^2) = AT$

3.9

First write S in terms of complex s.

$$S(s) = \frac{-s^2 + 1}{s^4 - 8s^2 + 16}$$

Next, find poles and zeros of $S(s)$

zeros: ± 1
poles: ± 2 (Both 2nd order)

Now use spectral factorization:

$$S(s) = \left[\frac{s+1}{(s+2)^2}\right]\left[\frac{-s+1}{(-s+2)^2}\right]$$

The first factor in the above expression
is the shaping filter; i.e.,

unity
white noise $\rightarrow$ $\boxed{\dfrac{s+1}{(s+2)^2}}$ $\rightarrow$ Noise with
spectral function $S(s)$

50

3.10

Transfer function:

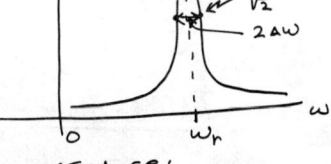

$$G(s) = \frac{R}{R + Ls + \frac{1}{Cs}}$$

$$= \frac{(R/L) s}{s^2 + \frac{R}{L} s + \frac{1}{LC}}$$

Now let $1/LC = w_r^2$ and $R/L = 2\zeta w_r$, and rewrite G:

$$G(s) = \frac{2\zeta w_r s}{s^2 + 2\zeta w_r s + w_r^2}$$

Noise equivalent bandwidth from Eq (3.5.3):

$$B = \frac{1}{2} \left[\frac{1}{2\pi j} \int_{-\infty}^{\infty} G(s) \cdot G(-s)\, ds \right] = \frac{1}{2} \zeta w_r \quad \begin{array}{l}\text{(From int.}\\ \text{tables on}\\ \text{p. 126)}\end{array}$$

Noise
Equiv. (Hz)

Next, find half-power bandwidth.

Replace s with $j\omega$:

$$G(j\omega) = \frac{2\zeta w_r j\omega}{(-\omega^2 + w_r^2) + j\omega 2\zeta w_r}$$

Check on normalization of frequency response:

Let $\omega = w_r$. Then

$$G(jw_r) = \frac{j 2\zeta w_r^2}{j 2\zeta w_r^2} = 1 \quad \text{(checks)}$$

Half power points occur where real = imag. in denominator (approximately)

$$\therefore \text{Let } 2\zeta w_r \omega = -\omega^2 + w_r^2$$

and solve for ω. Let solution be $w_r + \Delta\omega$. Substitute into above equation and obtain approximate $\Delta\omega$ by assuming ζ to be small. The result is $\Delta\omega = \zeta w_r \frac{rad}{sec}$

Therefore, half-power bandwidth in Hz is:

$$B_{(\frac{1}{2}\text{Power})} = \frac{2\zeta w_r}{2\pi} = \frac{\zeta w_r}{\pi} \text{ Hz. This compares with } \frac{\zeta w_r}{2} \text{ Hz for noise bandwidth.}$$

51

<u>3.11</u>

"Ideal" noise equivalent bandwidth $= \frac{1}{2\pi}$ H_3

(a) <u>First-order</u> Butterworth:

From Eq. 3.5.3

$$B = \frac{1}{2}\left[\frac{1}{2\pi j}\int_{-j\infty}^{j\infty}\frac{1}{s+1}\cdot\frac{1}{-s+1}\,ds\right]$$

Use tables on p. 126.

$$B = \frac{1}{2}\left[I_1\right] = \frac{1}{2}\frac{C_0}{2d_0d_1} = \frac{1}{2}\frac{1}{2\cdot1\cdot1} = \frac{1}{4}\ H_3$$

<u>Second-order</u> Butterworth:

$$B = \frac{1}{2}\left\{\frac{1}{2\pi j}\int_{-j\infty}^{j\infty}\frac{1}{s^2+\sqrt{2}s+1}\cdot\frac{1}{(-s)^2-\sqrt{2}s+1}\cdot ds\right\}$$

Use tables on p. 126.

$$B = \frac{1}{2}\frac{1}{2\cdot1\cdot\sqrt{2}} = \frac{1}{4\sqrt{2}} \approx \frac{1}{5.65}\ H_3$$

<u>Third-order</u> Butterworth:

$$B = \frac{1}{2}\left[\frac{1}{2\pi j}\int_{-j\infty}^{j\infty}\frac{1}{(s+1)(s^2+s+1)}\cdot\frac{1}{(-s+1)((-s)^2-s+1)}\cdot ds\right]$$

Use tables on p. 126. (Denom. $= s^3+2s^2+2s+1$)

$$B = \frac{1}{2}\cdot\frac{1\cdot2}{2\cdot1(2\cdot2-1)} = \frac{1}{6}\ H_3$$

(b) Third-order filter has noise B.W. of $\frac{1}{6}$ H_3 which is fairly close to the "ideal" filter with $\frac{1}{6.28}$ H_3. Therefore, increasing order of filter will not change noise output appreciably. However, increasing the order might be desirable from a signal fidelity viewpoint. This would not yield much more noise.

52

3.12

(a) $G(s) = \frac{1}{s^2}$; $g(t) = t$

$$E[x^2(t)] = \int_0^t \int_0^t g(u) g(v) A \delta(u-v) \, du \, dv \quad (Eq\ 3.8.4)$$

$$= \int_0^t \int_0^t u \cdot v \cdot A \cdot \delta(u-v) \, du \, dv = \int_0^t A v^2 \, dv$$

$$= A t^3/3$$

(b) $G(s) = \frac{1}{s^2 + \omega_0^2}$; $g(t) = \frac{1}{\omega_0} \sin \omega_0 t$

$$E[x^2(t)] = \int_0^t \int_0^t g(u) g(v) A \delta(u-v) \, du \, dv$$

$$= \int_0^t A g^2(v) \, dv = \frac{A}{\omega_0^2} \int_0^t \sin^2 \omega_0 v \, dv$$

$$= \frac{A}{\omega_0^3} \left[\frac{1}{2} \omega_0 t - \frac{1}{4} \sin 2\omega_0 t \right]$$

3.13

$$\ddot{x} + a\dot{x} = f(t), \quad or \quad s^2 X(s) + as X(s) = F(s)$$

Transfer fcn. $= G(s) = \frac{X(s)}{F(s)} = \frac{1}{s(s+a)}$

weighting fcn. $= \mathcal{L}^{-1}[G(s)] = g(t) = \frac{1}{a}(1 - e^{-at})$

$$E[x^2(t)] = \int_0^t \int_0^t g(u) g(v) A \delta(u-v) \, du \, dv = A \int_0^t g^2(v) \, dv$$

$$= A \int_0^t \frac{1}{a^2} [1 - e^{-av}]^2 \, dv$$

$$= \frac{A}{a^2} \left[t - \frac{2}{a}(1 - e^{-at}) + \frac{1}{2a}(1 - e^{-2at}) \right]$$

<u>3.14</u> This is a mixed deterministic /random problem so no special formulas should be used. Go back to basics. Superposition still applies, though.

$$G(s) = \frac{1}{1 + 10s} = \frac{1/10}{s + 1/10} \; ; \; g(t) = (1/10) e^{-t/10}$$

First, compute response due to $A u(t)$. Call it X_u.

$$X(s) = \frac{A \cdot 1/10}{s(s + 1/10)} \; ; \; \text{Therefore, } X_u(t) = A(1 - e^{-t/10})$$

Next, write response due to $n(t)$. Call it X_m

$$X_m(t) = \int_0^t g(\tau) n(t-\tau) d\tau \quad (n \text{ is unity w.m.})$$

Total response is then (call it $x(t)$):

$$X(t) = A(1 - e^{-t/10}) + \int_0^t g(\tau) m(t-\tau) d\tau$$

Next, compute <u>the mean</u>:

$$E(X(t)) = E\left[A(1 - e^{-t/10}) + \int_0^t g(\tau) m(t-\tau) d\tau \right]$$

$$= (1 - e^{-t/10}) \cdot E(A) + 0$$

A is uniform between 0 and 1, so $E(A) = 1/2$.

$$\therefore E[X(t)] = \frac{1}{2} (1 - e^{-t/10})$$

Next, compute <u>mean square</u> value:

$$E[X^2(t)] = E\left[A^2(1 - e^{-t/10})^2 + 2A(1 - e^{-t/10}) \int_0^t g(\tau) n(t-\tau) d\tau \right.$$
$$\left. + \int_0^t \int_0^t g(u) g(v) n(t-u) n(t-v) du dv \right]$$

$$= (1 - e^{-t/10})^2 \cdot E(A^2) + \int_0^t \int_0^t g(u) g(v) R_m(u-v) du dv$$

3.14 (cont.)

Evaluating both terms yields:

$$E(A^2) = \int_0^1 A^2 \cdot 1 \, dA = \frac{1}{3}$$

$$\int_0^t\int_0^t (---) \, du \, dv = \int_0^t\int_0^t g(u) g(v) \delta(u-v) \, du \, dv = \frac{1}{20}(1 - e^{-t/5})$$

The _variance_ is then found as:

$$Var \, x = E[x^2(t)] - [E(x(t))]^2$$

(a) For $t = .1$:

$$E(x) = \frac{1}{2}(1 - e^{-.1/10}) \approx .005$$

$$E(x^2) = \frac{1}{3}(1 - e^{-.1/10})^2 + \frac{1}{20}(1 - e^{-.1/5})$$

$$\approx .333 \times 10^{-4} + 10 \times 10^{-4} = 10.333 \times 10^{-4}$$

$$Var \, x = 10.333 \times 10^{-4} - .25 \times 10^{-4} = 10.083 \times 10^{-4}$$

(b) For $t = \infty$:

$$E(x) = \frac{1}{2} = .5$$

$$E(x^2) = \frac{1}{3} + \frac{1}{20} \approx .38333$$

$$Var \, x = .38333 - (.5)^2 = .1333$$

Note: In this problem one must be careful to consider all averages as ensemble averages. The output process is _not_ ergodic, so ensemble averages are not the same as time averages!

59

3.15

$$\text{Ave. value} = \frac{1}{T}\int_0^T A(t)\,dt = \frac{1}{T}\int_0^T [a_0 + m(t)]\,dt$$

$$= a_0 + \frac{1}{T}\int_0^T m(t)\,dt$$

Second term is error term, and rms value is desired.

$$E\left[(Error)^2\right] = E\left[\frac{1}{T^2}\int_0^T\int_0^T m(u)m(v)\,du\,dv\right] = \frac{1}{T^2}\int_0^T\int_0^T R_m(u-v)\,du\,dv$$

$$= \frac{1}{T^2}\int_0^T\int_0^T \sigma^2 e^{-\beta|u-v|}\,du\,dv$$

Because of absolute magnitude of $u-v$, evaluate in upper triangular region and the multiply result by 2.

$$\therefore\ E\left[(Error)^2\right] = \frac{2}{T^2}\int_0^T\int_0^v \sigma^2 e^{\beta(u-v)}\,du\,dv$$

Carrying out integration and taking square root yields

$$RMS\ ERROR = \frac{\sqrt{2}\,\sigma}{\beta T}\left[\beta T - (1 - e^{-\beta T})\right]^{1/2}$$

3.16

$$x(t) = \int_0^t f(u)e^{-au}\,du, \text{ and } x^2(t) = \int_0^t\int_0^t e^{-au}e^{-av}f(u)f(v)\,du\,dv$$

$$E[x^2(t)] = \int_0^t\int_0^t e^{-au}e^{-av}E[f(u)f(v)]\,du\,dv$$

$$= \int_0^t\int_0^t e^{-au}e^{-av}R_f(u-v)\,du\,dv$$

$$= \int_0^t\int_0^t e^{-au}e^{-av}\sigma^2 e^{-\beta|u-v|}\,du\,dv$$

3.16 (Cont.)

Because of absolute magnitude of u-v, evaluate integral in upper triangular region and multiply by 2.

$$E\{x^2(t)\} = 2\sigma^2 \int_0^t \int_0^v e^{-au} e^{-av} e^{\beta(u-v)} \, du \, dv$$

Evaluating the integral leads to:

$$E\{x^2(t)\} = \frac{2\sigma^2}{\beta-a}\left[\frac{1}{2a}(1-e^{-2at}) - \frac{1}{a+\beta}(1-e^{-(a+\beta)t})\right]$$

3.17

Initial value is $N(0, \sigma^2)$

$f(t) \longrightarrow \boxed{\frac{1}{s}} \longrightarrow x(t)$

$f(t)$ is white noise with autocorrelation fcn. $A\delta(\tau)$.

$$x(t) = x(0) + \int_0^t f(u) \, du, \qquad x(0) \text{ is } N(0, \sigma^2)$$

$$x^2(t) = x^2(0) + 2x(0)\int_0^t f(u) \, du + \int_0^t \int_0^t f(u) f(v) \, du \, dv$$

$$E\{x^2(t)\} = E\{x^2(0)\} + E\left[2x(0)\int_0^t f(u) \, du\right] + \int_0^t \int_0^t E\{f(u)f(v)\} \, du \, dv$$

$$= \sigma^2 + 0 + \int_0^t \int_0^t R_f(u-v) \, du \, dv$$

$$= \sigma^2 + \int_0^t \int_0^t A\delta(u-v) \, du \, dv$$

$$= \sigma^2 + At$$

3.18

$\overset{i.c.=0}{\underset{}{}} \quad \overset{i.c.=2}{\underset{}{}}$

$f(t) \longrightarrow \boxed{\frac{1}{S}} \longrightarrow \boxed{\frac{1}{S}} \longrightarrow x(t)$ $f(t)$ is unity white noise.

Transfer fcn from f to x is $G(s) = \frac{1}{s^2}$; or $g(t) = t$

$$x(t) = 2 + \int_0^t g(\tau) f(t-\tau) d\tau$$

$$x^2(t) = 4 + 4\int_0^t g(\tau) f(t-\tau) d\tau + \int_0^t \int_0^t g(u)g(v) f(t-u)f(t-v) du dv$$

$$E[x^2(t)] = 4 + 0 + \int_0^t \int_0^t g(u)g(v) R_f(u-v) du dv$$

$$= 4 + \int_0^t \int_0^t u v \, \delta(u-v) du dv = 4 + \int_0^t v^2 dv$$

$$= 4 + t^3/3$$

(a) At $t = 2$

$\quad E[x^2] = 4 + 8/3 = \frac{20}{3}$

$\quad E[x] = 2$

$p(x)$, $\sigma = \sqrt{8/3}$, 2, x

$Var\, x = \frac{20}{3} - 2^2 = 8/3$

(b) Prob. density is shown above.

3.19

This problem is similar to Example 3.9, p. 135.

Note that $g(t) = t$ and rewrite Eq (3.8.9) as

$$R_x(t_1, t_2) = \int_0^{t_2} \int_0^{t_1} u \cdot v \cdot \delta(u-v+t_2-t_1) du dv$$

Interchange order of integration:

$$R_x(t_1, t_2) = \int_0^{t_1} u \int_0^{t_2} v \cdot \delta(u-v+t_2-t_1) dv du$$

3.19 (con't.)

Now evaluated double integral. (For $t_2 > t_1$)

$$R_x(t_1, t_2) = \int_0^{t_1} u(u + t_2 - t_1) du = \frac{t_2 t_1^2}{2} - \frac{t_1^3}{6}, \quad t_2 > t_1$$

By symmetry we have for $t_2 < t_1$:

$$R_x(t_1, t_2) = \frac{t_1 t_2^2}{2} - \frac{t_2^3}{6}, \quad t_2 < t_1$$

3.20

Filter weighting function is shown at right.

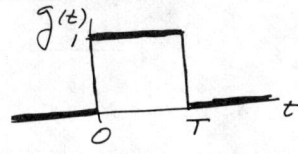

$$E[x^2(t)] = \int_0^t \int_0^t g(u) g(v) \delta(u - v) \, du \, dv$$

(a) For $0 < t < T$, weighting function is 1.

$$\therefore E[x^2(t)] = \int_0^t \int_0^t \delta(u - v) \, du \, dv = t$$

(b) For $t > T$ refer to figure at right. Contribution to to integral outside the $T \times T$ square is zero. Therefore, For $t > T$

$$E[x^2(t)] = T$$

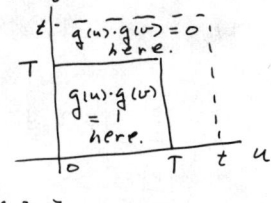

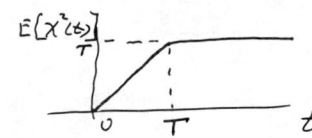

(c) Filter's memory is finite. The filter reaches the steady-state condition after an elapsed time of T units.

59

3.21

Use superposition and first work out transfer functions from input f_1 to x_1, x_2, x_3 and from input f_2 to x_1, x_2, x_3.

Consider input at f_1 (and $f_2 = 0$). Reduce inner loop with the result:

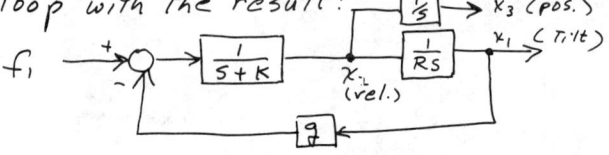

The respective transfer functions are: $\left(\begin{array}{l} 1^{st}\text{ subscript} = \text{Input} \\ 2^{nd}\text{ subscript} = \text{Output} \end{array}\right)$

$$G_{11}(s) = \frac{\frac{1}{(s+K)\,Rs}}{1 + \frac{1}{(s+K)\,Rs}\cdot g} = \frac{1/R}{s^2 + Ks + g/R} \qquad \text{(Tilt)}$$

$$G_{12}(s) = Rs \cdot G_{11}(s) = \frac{s}{s^2 + Ks + g/R} \qquad \text{(vel.)}$$

$$G_{13}(s) = \frac{1}{s} \cdot G_{12}(s) = \frac{1}{s^2 + Ks + g/R} \qquad \text{(Pos.)}$$

Next, consider the input at f_2 (and $f_1 = 0$). The block diagram is:

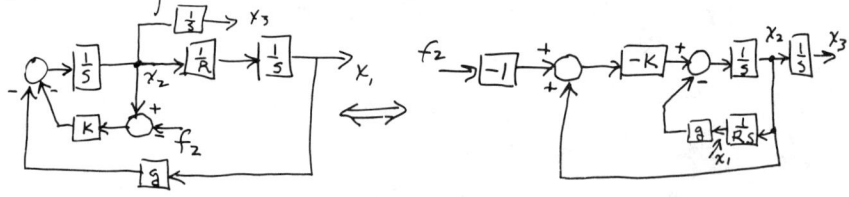

In figure to right, reduce inner loop first. This yields

$$\text{Inner loop} = \frac{\frac{1}{s}}{1 + \frac{1}{s}\cdot\frac{g/R}{s}} = \frac{s}{s^2 + g/R};$$

3.21 (con't.)

Respective transfer functions are:

$$G_{22}(s) = -\frac{\frac{-Ks}{s^2 + \vartheta/R}}{1 + \frac{Ks}{s^2 + \vartheta/R}} = \frac{Ks}{s^2 + Ks + \vartheta/R} \qquad (vel.)$$

$$G_{23}(s) = \frac{1}{s} \cdot G_{22}(s) = \frac{K}{s^2 + Ks + \vartheta/R} \qquad (pos.)$$

$$G_{21}(s) = \frac{1}{Rs} \cdot G_{22}(s) = \frac{K/R}{s^2 + Ks + \vartheta/R} \qquad (Tilt)$$

The respective mean square errors are found from:

$$E(x_1^2) = \frac{1}{2\pi j}\int_{-j\infty}^{j\infty} A_1 G_{11}(s)G_{11}(-s)\,ds + \frac{1}{2\pi j}\int_{-j\infty}^{j\infty} A_2 G_{21}(s)G_{21}(-s)\,ds$$

$$E(x_2^2) = \frac{1}{2\pi j}\int_{-j\infty}^{j\infty} A_1 G_{12}(s)G_{12}(-s)\,ds + \frac{1}{2\pi j}\int_{-j\infty}^{j\infty} A_2 G_{22}(s)G_{22}(-s)\,ds$$

$$E(x_3^2) = \frac{1}{2\pi j}\int_{-j\infty}^{j\infty} A_1 G_{13}(s)G_{13}(-s)\,ds + \frac{1}{2\pi j}\int_{-j\infty}^{j\infty} A_2 G_{23}(s)G_{23}(-s)\,ds$$

It is now routine to evaluate integrals using Table 3.1, p 126.

Example:
$$\frac{1}{2\pi j}\int_{-j\infty}^{j\infty} A_1 G_{11}(s)G_{11}(-s)\,ds = \frac{1}{2\pi j}\int_{-j\infty}^{j\infty} \frac{\sqrt{A_1}/R}{s^2 + Ks + \vartheta/R} \cdot \frac{\sqrt{A_1}/R}{(-s)^2 - Ks + \vartheta/R}\,ds$$

$$m = 2, \quad \begin{aligned} C_1 &= 0 \\ C_0 &= \sqrt{A_1}/R \end{aligned} \quad, \quad \begin{aligned} d_2 &= 1 \\ d_1 &= K \\ d_0 &= \vartheta/R \end{aligned} \quad, \quad I_2 = \frac{C_0^2 d_2}{2 d_0 d_1 d_2} = \frac{A_1/R^2}{2\cdot\frac{\vartheta}{R}\cdot K\cdot 1} = \frac{A_1}{2KR\vartheta}$$

Final result:

$$E(x_1^2) = \frac{A_1 + A_2 K^2}{2KR\vartheta} \qquad (Tilt)$$

$$E(x_2^2) = \frac{A_1/2K + \frac{KA_2}{2}}{} \qquad (vel.)$$

$$E(x_3^2) = \frac{A_1 + A_2 K^2}{2K\vartheta/R} \qquad (pos.)$$

3.22

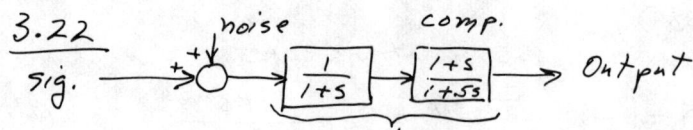

(a) The compensator opens up the bandwidth and thus frequency response is flatter within bandwidth of signal. Thus, looking at signal alone, it appears signal fidelity is improved by insertion of the compensator.

(b) Before insertion of compensator:

$$S = \text{Sig. Pwr.} = \frac{1}{2\pi}\int_{-1}^{1}\frac{1}{1+\omega^2}d\omega = \frac{2}{2\pi}\cdot\tan^{-1}1 = .25$$

$$N = \text{Noise Pwr.} = \frac{1}{2\pi}\int_{-\infty}^{\infty}.1\frac{1}{1+\omega^2}d\omega = .1\frac{2}{2\pi}\tan^{-1}\infty = .05$$

$$\therefore S/N = 5$$

After insertion of compensator:

$$S = \text{Sig. Pwr.} = \frac{1}{2\pi}\int_{-1}^{1}\frac{1}{1+\omega^2/4}d\omega \approx .295$$

$$N = \text{Noise Pwr.} = \frac{1}{2\pi}\int_{-\infty}^{\infty}.1\frac{1}{1+\omega^2/4}d\omega = 0.1$$

$$\therefore S/N = 2.95$$

(c) Insertion of compensator results in a reduction of signal-to-noise ratio from 5 to 2.95. In effect, opening up the bandwidth yields a modest improvement in signal transmission, but the price is high! Twice as much noise is passed as without the compensator.

3.23

$f(t) \xrightarrow{+} \bigcirc \xrightarrow{+} \boxed{G(s)} \longrightarrow X(t)$

$m(t) \longrightarrow \boxed{\text{Cross-Corr.}} \longrightarrow R_{mx}(\nu)$

From Fourier transform theory:

$$X(t) = \int_{-\infty}^{\infty} g(u)\left[f(t-u) + m(t-u)\right] du$$

Cross correlator output:

$$R_{mx}(\nu) = E\left[m(t) \, x(t+\nu)\right]$$

$$= E\left[m(t) \int_{-\infty}^{\infty} g(u)\left[f(t+\nu-u) + m(t+\nu-u)\right] du\right]$$

$$= \int_{-\infty}^{\infty} g(u) \cdot E\left[m(t) \cdot m(t+\nu-u)\right] du$$

$$= \int_{-\infty}^{\infty} g(u) \, K \, \delta(\nu-u) \, du = K g(\nu)$$

There are many subtleties in implementation
of this scheme. Most of them trace back to the
expectation operator in above derivation. This
calls for ensemble averaging over an infinite
number of member functions of the process.
Usually, one exchanges ensemble averaging for
time averaging, and under these circumstances
the above is best implemented using FFT
methods in the frequency domain (with all its
pitfalls). For example:

$$\mathcal{F}[R_{mx}(\nu)] \approx \mathcal{F}\left[\frac{1}{T} \int_0^T m(u) x(u+\nu) \, du\right]$$

$= N_T^*(j\omega) \cdot X_T(j\omega)$, where N_T and X_T are Fourier trans. of
truncated m and x.

3.24

(a)   Ignore the noise for the moment.

Trans. fcn. $= \dfrac{Y(s)}{\theta_i(s)}$

$$= \frac{K_d}{1 + K_d \cdot \dfrac{K_o}{s}} = \frac{K_d\, s}{s + K_d K_o}$$

This can also be written as:

$$s: \quad \frac{K_d}{s + K_d K_o} \quad ; \quad \text{Block diagram:}$$

(b)   Transfer function for noise is:

$$\frac{Y(s)}{N(s)} = \frac{K_d}{1 + K_d \cdot \dfrac{K_o}{s}} = \frac{K_d\, s}{s + K_d K_o}$$

Now assume cutoff frequency of the filter, $K_d K_o$, is large. Then the transfer characteristic to the noise is approximately

$$\frac{K_d (j\omega)}{j\omega + K_d K_o} \approx \frac{j\omega}{K_o} \quad (\text{for small } \omega)$$

Denote output noise as $n'(t)$. Its spectral function is then

$$S_{n'} \approx N_o \left| \frac{j\omega}{K_o} \right|^2 = \left(N_o / K_o^2 \right) \omega^2$$

$$\text{Noise Power} = E\{(n')^2\} = \frac{1}{2\pi} \int_{-2\pi W}^{2\pi W} N_o / K_o^2 \, \omega^2 d\omega = \frac{2}{3} \frac{N_o}{K_o^2} (2\pi)^2 W^3$$

3.25

$$\gamma_{xy}^2(\omega) = \frac{|S_{xy}(j\omega)|^2}{S_x(j\omega) \cdot S_y(j\omega)} \quad ; \qquad x \longrightarrow \boxed{G(\omega)} \longrightarrow y$$

(a) $y(t) = \int_{-\infty}^{\infty} g(u) x(t-u)\, du$

$$S_{xy}(j\omega) = \mathcal{F}\{R_{xy}(\tau)\} = \mathcal{F}\left[E\left[x(t) y(t+\tau) \right] \right]$$

$$= \mathcal{F}\left[E\left[x(t) \int_{-\infty}^{\infty} g(u) x(t+\tau-u)\, du \right] \right]$$

$$= \mathcal{F}\left[\int_{-\infty}^{\infty} g(u)\, E\left[x(t) x(t+\tau-u) \right] du \right]$$

$$= \mathcal{F}\left[\int_{-\infty}^{\infty} g(u) R_x(\tau-u) \right] = G(j\omega) S_x(j\omega)$$

(b) $\gamma_{xy}^2 = \dfrac{|G(j\omega)|^2 S_x(j\omega) \cdot S_x(j\omega)}{S_x(j\omega) \cdot \underbrace{S_x(j\omega)\,|G(j\omega)|^2}_{S_y(j\omega)}}$ (Note S_x is Real.)

$$= 1$$

CHAPTER 4

<u>4.1</u>   Use Eq. (4.2.5) and integral tables on p.126.

$$G(s) = \frac{1}{(s/w_c)^2 + \sqrt{2}\,(s/w_c) + 1} \;;\; 1 - G(s) = \frac{(s/w_c)^2 + \sqrt{2}\,(s/w_c)}{(s/w_c)^2 + \sqrt{2}(s/w_c) + 1}$$

$$E(e^2) = \frac{1}{2\pi j} \int_{-j\infty}^{j\infty} \frac{\sqrt{2}}{(s+1)} \cdot \frac{[(s/w_c)^2 + \sqrt{2}\,(s/w_c)]}{(s/w_c)^2 + \sqrt{2}\,(s/w_c) + 1} \cdot \begin{bmatrix} \text{mirror image} \\ \text{in right plane} \end{bmatrix} ds$$

$$+ \frac{1}{2\pi j} \int_{-j\infty}^{j\infty} \frac{1}{(s/w_c)^2 + \sqrt{2}\,(s/w_c) + 1} \cdot \begin{bmatrix} \text{Mirror} \\ \text{Image} \end{bmatrix} ds$$

The above expression is then evaluated using integral tables. The result is:

$$E(e^2) = \frac{\sqrt{2} + 3\,w_c}{\sqrt{2} + 2w_c + \sqrt{2}\,w_c^2} + \frac{w_c}{2\sqrt{2}}$$

This can be evaluated numerically for a few values of w_c between 0 and 1. The results:

w_c	$E(e^2)$
1	1.268
.5	1.230
.25	1.169
.125	1.105
$\vdots$	$\vdots$
0	1.000 $\leftarrow$ Minimum

The minimum is 1.0 and occurs at $w_c = 0$; i.e., this is a "no-pass" filter.

This strange result occurs because of the relatively large noise and the constraint on the shape of the response in the pass. Had we allowed a more general 2nd order form, say,

$$G(s) = \frac{w_0^2}{s^2 + 2\xi w_0 s + w_0^2}$$

and optimized w.r.t. w_0 and ξ, we would have done better.

4.2

The servo in this case acts like a low pass filter, and the problem is to adjust K to give the least mean square error.

Use Eq. (4.2.5) where $G(s)$ is overall closed loop transfer function. In this case it is

$$G(s) = \frac{\frac{5K}{s(s+5)}}{1 + \frac{5K}{s(s+5)}} = \frac{5K}{s^2 + 5s + 5K}$$

and

$$1 - G(s) = \frac{s^2 + 5s}{s^2 + 5s + 5K}$$

Mean square error is then

$$E(e^2) = \frac{1}{2\pi j}\int_{-j\infty}^{j\infty} \frac{s^2 + 5s}{s^2 + 5s + 5K} \cdot \frac{(-s)^2 - 5s}{(-s)^2 - 5s + 5K} \cdot S_s(s)\, ds$$

$$+ \frac{1}{2\pi j}\int_{-j\infty}^{j\infty} \frac{5K}{s^2 + 5s + 5K} \cdot \frac{5K}{(-s)^2 - 5s + 5K} \cdot S_m(s)\, ds$$

Now, note that the first term can be written as

$$\frac{1}{2\pi j}\int_{-j\infty}^{j\infty} \frac{s+5}{s^2 + 5s + 5K} \cdot \frac{-s+5}{(-s)^2 - 5s + 5K} \cdot (s)(-s) S_s(s)\, ds$$

Now note that $(s)(-s)S_s(s)$ is P.S.D. of $\dot{s}(t)$. $E[e^2]$ is then

$$E[e^2] = \frac{1}{2\pi j}\int_{-j\infty}^{j\infty} \left[\frac{(s+5)}{s^2 + 5s + 5K} \cdot \frac{\sqrt{1000}}{s+1}\right]\left[\begin{array}{c}\text{mirror image in} \\ \text{Right half plane}\end{array}\right] ds$$

$$+ \frac{1}{2\pi j}\int_{-j\infty}^{j\infty} \left[\frac{5K}{s^2 + 5s + 5K} \cdot 1\right]\left[\begin{array}{c}\text{mirror} \\ \text{Image}\end{array}\right] ds$$

4.2 (cont.)

$E[e^2]$ can now be evaluated with int. tables on p. 126
The result is:

$$E[e^2] = \frac{1000}{K}\left[\frac{.1K + 25.5}{5K + .51}\right] + \frac{K}{2}$$

Relative minimum can now be found either numerically or by differential calculus. The result is:

$$K \approx 27$$

The corresponding damping ratio is obtained by noting the characteristic function is

$$s^2 + 5s + 135 \qquad (\text{std form is: } s^2 + 2\zeta\omega_o s + \omega_o^2)$$

Thus damping ratio is:

$$\text{Damping ratio} = \zeta = \frac{5}{2\sqrt{135}} \approx .215$$

4.3

Refer to the results of Problem 3.21.

$$E(x_3^2) = \frac{A_1}{2K\omega_o^2} + \frac{A_2 K}{2\omega_o^2}, \quad \text{where } \omega_o^2 = q/R$$

Use diff. calculus to find a relative minimum.

$$\frac{dE(x_3^2)}{dK} = -\frac{A_1}{2K^2\omega_o^2} + \frac{A_2}{2\omega_o^2} = 0$$

$$\text{or} \quad \frac{A_1}{K^2} = A_2$$

$$\text{Or} \quad K = \sqrt{\frac{A_1}{A_2}}$$

68

4.4.   First write the equation for relative
motion y (see figure in book). From Newton's law:

$$M(\ddot{x}-\ddot{y}) = Ky + B\dot{y}$$

OR $$\ddot{y} + B/m\,\dot{y} + \frac{K}{m}\,y = \ddot{x}$$

Now think of $\ddot{x}$ as input and y as response
and obtain transfer function relating the two.

$$\frac{s\,Y(s)}{s\,X(s)} = \frac{s^2}{s^2 + B/m\,s + K/m}$$

Now let

$$K/m = \omega_0^2$$

$$B/m = 2\xi\omega_0 = 2\omega_0 \quad \left(\begin{array}{l}\text{Because we}\\ \text{assume critical}\\ \text{damping as } K\\ \text{is varied.}\end{array}\right)$$

Therefore,

$$\frac{s\,Y(s)}{s\,X(s)} = \frac{s^2}{s^2 + 2\omega_0 s + \omega_0^2}, \quad \begin{array}{l}(\omega_0 \text{ is the}\\ \text{parameter to}\\ \text{be varied.})\end{array}$$

Next, look at average power.

Instantaneous Power = Force $\times$ velocity

$$= (B\dot{y}) \times \dot{y}$$

B is $2\omega_0 m$ because of critical damping.
Therefore, average power is:

$$P_{ave.} = E[2\omega_0 M\dot{y}\cdot\dot{y}] = 2\omega_0 M\cdot E(\dot{y}^2)$$

We need to evaluate mean square value of
$\dot{y}$, given the spectral density of $\ddot{x}$, which
is given; i.e.,

$$R_{\ddot{x}}(\tau) = \sigma^2 e^{-\beta|\tau|} \quad ; \quad S_{\ddot{x}}(s) = \frac{2\sigma^2\beta}{-s^2 + \beta^2}$$

69

4.4 (cont.)

$$E(\dot{y}^2) = \frac{1}{2\pi j} \int_{-j\infty}^{j\infty} \left[\frac{s^2}{s^2 + 2\omega_0 s + \omega_0^2} \cdot \frac{\sqrt{2\sigma^2\beta}}{s+\beta} \right] \left[\begin{array}{c} \text{Mirror image} \\ \text{in right half} \\ \text{plane} \end{array} \right] ds$$

$$= \frac{\sigma^2 \beta (\omega_0 + 2\beta)}{2 (\omega_0 + \beta)^2} \qquad \text{(From tables on p. 126.)}$$

Average power is therefore:

$$P_{ave.} = 2\omega_0 M \cdot E(\dot{y}^2) = \sigma^2\beta M \frac{\omega_0 (\omega_0 + 2\beta)}{(\omega_0 + \beta)^2}$$

This is to be maximized w.r.t. ω_0, which, in turn, is related to the spring constant K. A sketch of $P_{ave.}$ is shown below.

Approaches $\sigma^2\beta M$ as $\omega_0 \to \infty$.

Maximum occurs at $\omega_0 = \infty$, so we simply let ω_0 be "large" (∴ K is "large"); then we look for the upper limit of P_{ave} for large ω_0.

$$\text{Maximum } P_{ave.} = \sigma^2\beta M$$

For $\sigma^2 = 1 \, (\text{ft/sec})^2$, $\beta = 2\pi \, (\text{rad/sec})$, $M = .1 \, Kg$

$$\text{Maximum } P_{ave.} \approx 58 \, mw.$$

(Note: In first printing of the book, β was given as 2 rather than 2π. Long used 2π in his dissertation, so this value is preferred.)

4.5

$$S_s = \frac{8}{-s^2+4} \quad , \quad S_m = \frac{2}{-s^2+1}$$

$$S_{s+m} = S_s + S_m = \frac{-10s^2+16}{(-s^2+4)(-s^2+1)} = 10\frac{-s^2+\frac{8}{5}}{(-s^2+4)(-s^2+1)}$$

$$\frac{S_s}{S_{s+m}^-} = \frac{\frac{8}{(s+2)(-s+2)}}{\sqrt{10}\frac{-s+\sqrt{8/5}}{(-s+2)(-s+1)}} = \frac{8}{\sqrt{10}}\frac{-s+1}{(s+2)(-s+\sqrt{8/5})}$$

$$= \frac{8}{\sqrt{10}}\left[\frac{K_1}{s+2} + \frac{K_2}{-s+\sqrt{8/5}}\right] , \quad K_1 = \frac{3}{2+\sqrt{8/5}}$$

$$K_2 \text{ is not needed.}$$

Optimum $G(s) = \dfrac{1}{\sqrt{10}\dfrac{s+\sqrt{8/5}}{(s+2)(s+1)}}\left[\dfrac{8}{\sqrt{10}}\cdot\dfrac{3}{2+\sqrt{8/5}}\cdot\dfrac{1}{s+2}\right]$

$$= \frac{12}{5}\cdot\frac{1}{2+\sqrt{8/5}}\cdot\frac{s+1}{s+\sqrt{8/5}}$$

To find mean square error, use Eq (4.3.12):

$$E(e^2) = R_s(0) - \int_0^\infty g(u)\, R_s(u)\, du$$

In this case,

$$R_s(0) = 2, \quad g(u) = \frac{12}{5}\cdot\frac{1}{2+\sqrt{8/5}}\cdot\left[\delta(u) + (1-\sqrt{8/5})\,e^{-\sqrt{8/5}\,u}\right],$$

and $R_s(u) = 2e^{-2u}$ (for $u > 0$)

Substituting the above into Eq (4.3.12) yields

$$E(e^2) \approx .65$$

(Note: See solution to Prob. 4.6 for a
comment on this result.)

4.6 Noncausal solution for Prob. 4.5. As before:

$$S_s = \frac{8}{-s^2+4} \; , \quad S_m = \frac{2}{-s^2+1} \; , \quad S_{s+m} = \frac{-10s^2+16}{(-s^2+4)(-s^2+1)}$$

$$\text{Optimal noncausal } G(s) = \frac{S_s}{S_{s+m}} = \frac{8}{10} \cdot \frac{-s^2+1}{-s^2+\tfrac{8}{5}}$$

Weighting function $g(u)$ is found by rewriting the above $G(s)$ as

$$G(s) = \frac{8}{10}\left[1 - \frac{-.6}{-s^2+\tfrac{8}{5}} \right]$$

The inverse 2-sided Laplace transform will be recognized as

$$g(u) = .8\, \delta(u) - \frac{.48}{2\sqrt{\tfrac{8}{5}}}\, e^{-\sqrt{\tfrac{8}{5}}\,|u|}$$

Mean square error is found from Eq (4.3.12)

This leads to: $\quad E(e^2) \approx .63$

Note that this is only slightly less than the .65 value obtained in the causal case (Prob 4.5). This is due to the "convenient" numbers used in this example. The noise and the signal have almost the same spectral characteristics, so the resulting optimal filter is almost flat with frequency. As a matter of fact, if we were to consider a trivial flat-gain filter, the best gain works out to be $\frac{2}{3}$ and its $E[e^2]$ is $\frac{2}{3}$, only slightly more than .63 or .65.

<u>4.7</u> $S_s = \dfrac{\omega^2+1}{\omega^4+8\omega^2+16}$; or $S_s(s) = \dfrac{-s^2+1}{s^4+8(-s^2)+16}$

Assume zero noise. Therefore $S_{s+m} = S_s$

First, factor S_{s+m} :

$$S_{s+m} = S_{s+m}^+ \cdot S_{s+m}^- = \frac{s+1}{(s+2)(s+2)} \cdot \frac{-s+1}{(-s+2)(-s+2)}$$

Next, find S_s/S_{s+m}^- and its inverse.

$$S_s/S_{s+m}^- = \frac{s+1}{(s+2)(s+2)} = \left[\frac{-1}{(s+2)^2} + \frac{1}{(s+2)} \right]$$

$$\mathscr{L}^{-1}\left[S_s/S_{s+m}^-\right] = -t\,e^{-2t} + e^{-2t},$$

Now shift to the left
one unit as shown.
Call the shifted function $g_{sh}(t)$

$$g_{sh}(t) = \begin{cases} -(t+1)e^{-2(t+1)} + e^{-2(t+1)} & , \quad t > -1 \\ 0 & , \quad t < -1 \end{cases}$$

Now truncate the negative-time part and take the ordinary single-sided L.T. of the result. Truncated fcn. for $t>0$ is $-e^{-2}t\,e^{-2t}$. Its Laplace transform is $-e^{-2}\dfrac{1}{(s+2)^2}$. Therefore, Optimal predictor $G(s)$ is:

$$G(s) = \frac{-e^{-2}\frac{1}{(s+2)^2}}{\frac{(s+1)}{(s+2)^2}} = -e^{-2}\frac{1}{s+1}$$

<u>4.8</u>

$$S_s = \frac{2}{-s^2+1} \quad , \quad S_m = \frac{4}{-s^2+4}$$

$$S_{s+m} = S_s + S_m = \frac{-6s^2 + 12}{(-s^2+1)(-s^2+4)}$$

(a) Optimal noncausal solution:

$$G(s) = \frac{S_s}{S_{s+m}} = \frac{1}{3}\frac{-s^2+4}{-s^2+2}$$

or

$$G(s) = \frac{1}{3}\left[1 + \frac{2}{-s^2+2}\right]$$

By inspection, the weighting function is then

$$g(u) = \frac{1}{3}\delta(u) + \frac{1}{3}\cdot\frac{1}{\sqrt{2}} e^{-\sqrt{2}|u|}$$

(b) Yes. Consider the problem of estimating the signal somewhere in the interior region of a long span of noisy measurement data, ie., the off-line smoothing problem.

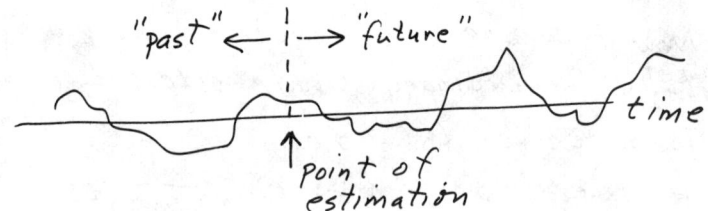

"past" ← | → "future"

time

↑ point of estimation

Measurement data is available both ahead and behind the point of interest, so a non-causal weighting function may be used. The calculated $E(e^2)$ is just the smoothing mean sq. error.

74

4.9 $\quad R_s(\tau) = 4e^{-4|\tau|}$, $\quad S_s = \dfrac{2 \cdot 4 \cdot 4}{-s^2 + 16}$

$\quad R_m(\tau) = e^{-|\tau|}$, $\quad S_m = \dfrac{2 \cdot 1 \cdot 1}{-s^2 + 1}$

$$S_{s+m} = S_s + S_m = 34 \frac{-s^2 + \frac{32}{17}}{(-s^2 + 16)(-s^2 + 1)}$$

Next, find S_s / S_{s+m}^- and its inverse.

$$\frac{S_s}{S_{s+m}^-} = \frac{\dfrac{32}{(s+4)(-s+4)}}{\sqrt{34} \ \dfrac{-s + \sqrt{\frac{32}{17}}}{(-s+4)(-s+1)}} = \frac{32}{\sqrt{34}} \ \frac{-s+1}{(s+4)(-s+\sqrt{\frac{32}{17}})}$$

$$= \frac{32}{\sqrt{34}} \left[\frac{K_1}{s+4} + \frac{K_2}{-s+\sqrt{\frac{32}{17}}} \right] , \quad K_1 = \frac{5}{4 + \sqrt{\frac{32}{17}}}$$

Only the positive-time part of the inverse is of interest because we shift to the left in prediction problem. Therefore, the unshifted inverse is:

$$\frac{32}{\sqrt{34}} K_1 e^{-4t} + (\text{neg. time term})$$

A shift to the left of .25 units replaces t with $t + .25$. After truncating the neg. time part we have

$$\text{Pos.-time Part (after shift)} = e^{-1} \frac{32}{\sqrt{34}} K_1 e^{-4t}$$

In the s domain, this is $e^{-1} \dfrac{32}{\sqrt{34}} \cdot K_1 \cdot \dfrac{1}{s+4}$

Finally the optimal transfer function is

$$G(s) = \frac{e^{-1} \dfrac{32}{\sqrt{34}} K_1 \dfrac{1}{s+4}}{S_{s+m}^+} = \frac{e^{-1} \dfrac{32}{\sqrt{34}} K_1 \dfrac{1}{s+4}}{\sqrt{34} \ \dfrac{s + \sqrt{32/17}}{(s+4)(s+1)}}$$

$$= e^{-1} \frac{32}{34} \cdot \frac{5}{4 + \sqrt{32/17}} \cdot \frac{s+1}{s + \sqrt{32/17}}$$

4.10 Differentiate Eq. (4.3.6) twice:

$$\frac{d\,E(\epsilon^2)}{d\epsilon} = \int_{-\infty}^{\infty}\int_{-\infty}^{\infty}\left[\eta(u)g(v) + g(u)\eta(v) + 2\epsilon\,\eta(u)\eta(v)\right]R_{s+m}(u-v)\,du\,dv$$

$$- 2\int_{-\infty}^{\infty}\eta(u)\,R_{s+m,s}(\alpha+u)\,du$$

$$\frac{d^2 E(\epsilon^2)}{d\epsilon^2} = 2\int_{-\infty}^{\infty}\int_{-\infty}^{\infty}\eta(u)\,\eta(v)\,R_{s+m}(u-v)\,du\,dv$$

The above double integral will be recognized as the formula for the mean square output of a filter $\eta(u)$ with $s+n$ as its input. This must be positive. Therefore, the extremum must be a minimum.

4.11

Assume $E[e(t)\cdot \mathfrak{z}(v)] = 0$, for $\bar{o} < v < t$

where $\mathfrak{z} = $ Input $= s + m$

But,

$$e(t) = s(t+\alpha) - \hat{s}(t+\alpha) = s(t+\alpha) - \int_0^t g(u)\,\mathfrak{z}(t-u)\,du$$

Substituting for $e(t)$ in above yields

$$E[e(t)\cdot \mathfrak{z}(v)] = E\left\{\left[s(t+\alpha) - \int_0^t g(u)\,\mathfrak{z}(t-u)\,du\right]\cdot\mathfrak{z}(v)\right\}$$

$$= E[\mathfrak{z}(v)s(t+\alpha)] - E\left[\int_0^t g(u)\,\mathfrak{z}(v)\,\mathfrak{z}(t-u)\,du\right]$$

$$= R_{s+m,s}(t-v+\alpha) - \int_0^t g(u)\,R_{s+m}(t-v-u)\,du$$

Now, orthog. principle says above is zero for $o < v < t$. Next, let $t - v = \tau$ and think of t as fixed. Then for $o < v < t,\ o < \tau < t$. The above is then equivalent to:

$$R_{s+m,s}(\tau+\alpha) - \int_0^t g(u)\,R_{s+m}(\tau-u)\,du = 0,\quad o < \tau < t$$

and this is what we set out to show.

76

<u>4.12</u>   Approximate the signal as a Markov process with a large time constant. Therefore,

$$R_s(\tau) = \sigma^2 e^{-\beta|\tau|}, \quad S_s(s) = \frac{2\sigma^2\beta}{-s^2+\beta^2}, \quad \beta \ll 1$$

Noise is white, so $S_n = A$

$$\therefore S_{s+n} = \frac{2\sigma^2\beta}{-s^2+\beta^2} + A = \frac{A(-s^2) + 2\sigma^2\beta + A\beta^2}{-s^2+\beta^2}$$

Eq. (4.4.8) gives the diff. eq. for the weighting fcn.

$$A\left(-\frac{d^2}{d\tau^2}\right)g(\tau) + (2\sigma^2\beta+A\beta^2)g(\tau) = \left(-\frac{d^2}{d\tau^2}+\beta^2\right)e^{-\beta\tau}$$

Right side of above is zero, so the solution is:

$$g(\tau) = C_1 e^{\sqrt{\frac{2\sigma^2\beta}{A}+\beta^2}\,\tau} + C_2 e^{-\sqrt{\frac{2\sigma^2\beta}{A}+\beta^2}\,\tau}$$

Now note β is small so solution is approx. constant for reasonable range of τ. Therefore, let $g(\tau) \approx k$ and substitute into integral eq.

$$\int_0^t k\, R_{s+n}(\tau-u)\, du = \sigma^2 e^{-\beta\tau} \approx \sigma^2$$

or

$$k\int_0^t \left[\sigma^2 + A\delta(\tau-u)\right] du \approx \sigma^2$$

$$k\sigma^2 t + A k \approx \sigma^2$$

or

$$k \approx \frac{\sigma^2}{\sigma^2 t + A} = \frac{1}{t + \left(\frac{A}{\sigma^2}\right)}, \quad \begin{array}{l} \text{where} \\ \sigma^2 = \text{Var of } a_0 \\ A = \text{white} \\ \quad\text{noise ampl.} \end{array}$$

Sketch of wt. fcn.

$g(\tau)$, $\frac{1}{t+\frac{A}{\sigma^2}}$, τ (age variable)

<u>4.13</u> $\quad g(\tau) = a(t)\, e^{-\sqrt{3}\,\tau} + b(t)\, e^{\sqrt{3}\,t}$

Substitute above into Eq (4.4.4) and evaluate.

$R_{s+m,s} = R_s = e^{-\tau}$, for $\tau > 0$; $\quad R_{s+m} = \delta(\tau) + e^{-|\tau|}$

Left side of integral equation is

$$\int_0^t g(u)\left[\delta(\tau-u) + e^{-|\tau-u|}\right] du$$

$$= g(\tau) + \int_0^\tau g(u)\, e^{-(\tau-u)}\, du + \int_\tau^t g(u)\, e^{(\tau-u)}\, du$$

Now substitute above expression for g and rewrite as

$$g(\tau) + e^{-\tau} a(t) \int_0^\tau e^{(1-\sqrt{3})u}\, du \;+\; e^{-\tau} b(t) \int_0^\tau e^{(1+\sqrt{3})u}\, du$$

$$+\; e^{\tau} a(t) \int_\tau^t e^{-(1+\sqrt{3})u}\, du \;+\; e^{\tau} b(t) \int_\tau^t e^{-(1-\sqrt{3})u}\, du$$

Evaluate the integrals next. It is best to leave $a(t)$ and $b(t)$ in general form at this point. After evaluating integrals, collect terms in $e^{-\sqrt{3}\,\tau}$ and $e^{\sqrt{3}\,\tau}$. These terms must vanish because right side of eq. contains only a term of the form $e^{-\tau}$. This "checks":

$$a(t)\left[1 + \frac{1}{1-\sqrt{3}} - \frac{1}{(1+\sqrt{3})}\right] e^{-\sqrt{3}\,\tau} = 0; \quad b(t)\left[1 + \frac{1}{1+\sqrt{3}} - \frac{1}{\sqrt{3}-1}\right] e^{\sqrt{3}\,\tau} = 0$$

Next, collect terms in $e^{-\tau}$ and e^{τ}. Equating like terms on both sides of integral equation leads to 2 equations of the form:

$$a(t)\{\ \ \} + b(t)\{\ \ \} = 1$$

$$a(t)\{\ \ \} + b(t)\{\ \ \} = 0$$

Now solve these equations for $a(t)$ and $b(t)$ to get values given by Eqs (4.4.14) and (4.4.17).

<u>4.14</u> Look at accelerometer channel first. Redraw figure so as to refer input to the h level rather than $\ddot{h}$

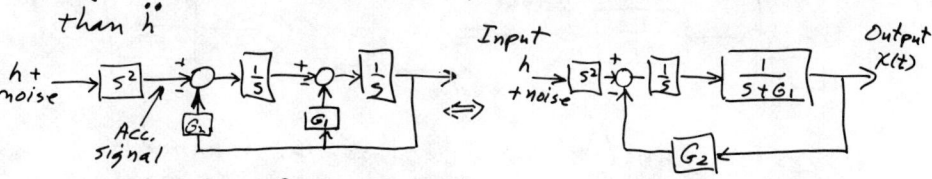

Transfer fcn. from h input to x output = $Y_a(s)$

$$Y_a(s) = \frac{\frac{1}{s(s+G_1)}}{1 + \frac{1}{s(s+G_1)} G_2} \cdot s^2 = \frac{s^2}{s^2 + G_1 s + G_2}$$

Next, look at barometric channel.

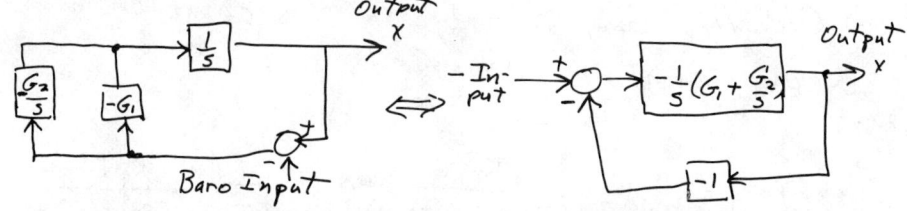

Transfer function from input to output = $Y_b(s)$.

$$Y_b(s) = - \frac{-\frac{1}{s}(G_1 + \frac{G_2}{s})}{1 + \frac{1}{s}(G_1 + \frac{G_2}{s})} = \frac{G_1 s + G_2}{s^2 + G_1 s + G_2}$$

(a) Note that $Y_a + Y_b = 1$ and the complementary constraint is satisfied.

(b) Clearly, from the form of the denominator (both Y_a and Y_b), the system is second order and G_1 and G_2 may be chosen to yield the desired nat. frequency and damping ratio.

4.15

Tach

Transfer function for tachometer channel:
$$G_T(s) = \frac{1}{1 + Ts}$$

Transfer function for accelerometer channel (referred to velocity level):
$$G_A(s) = \frac{Ts}{1 + Ts}$$

(a) Clearly, $G_T + G_A = 1$, so the complementary constraint is satisfied.

(b)
$$E[e^2] = \underbrace{\frac{1}{2\pi j}\int_{-j\infty}^{j\infty} \frac{K_T}{(Ts+1)(-Ts+1)}\,ds}_{I_1} + \underbrace{\frac{1}{2\pi j}\int_{-j\infty}^{j\infty} \frac{K_A T^2}{(Ts+1)(-Ts+1)}\,ds}_{I_1'}$$

use int.
Tables on
p. 126.

$c_0 = \sqrt{K_T}, \quad d_1 = T$
$d_0 = 1$

$c_0' = \sqrt{K_A}\,T, \quad d_1' = T$
$d_0' = 1$

$I_1 = \frac{c_0^2}{2d_0 d_1} = \frac{K_T}{2T}$

$I_1' = \frac{c_0'^2}{2d_0' d_1'} = \frac{K_A T^2}{2T}$

Mean square error is then:
$$E(e^2) = \frac{K_T}{2T} + \frac{K_A T}{2}$$

Differentiate w.r.t. T and set equal to zero:
$$dE(e^2)/dT = -\frac{K_T}{2T^2} + \frac{K_A}{2} = 0$$

Or, $\quad T = \sqrt{K_T/K_A}$

4.16   Complementary filter:

$$\begin{matrix} \text{Output} \\ \text{(in s-} \\ \text{domain)} \end{matrix} = S + \underbrace{N_1 G_1 + N_2 G_2 + \cdots N_N (1 - G_1 - G_2 \cdots)}_{\text{Error Term}}$$

General N-1 Input Wiener Problem:

$x =$ Signal

$\left.\begin{matrix} m_1 \\ m_2 \\ m_3 \\ \vdots \end{matrix}\right\} =$ noises

$$\begin{matrix} \text{Filter error} \\ \text{(in s-domain)} \end{matrix} = X(s) - \hat{X}(s)$$

$$= X - \left[(X + M_1) Y_1 + (X + M_2) Y_2 + \cdots \right]$$

$$= -M_1 Y_1 - M_2 Y_2 \cdots - M_{N-1} Y_{N-1}$$

$$+ X \left[1 - Y_1 - Y_2 \cdots - Y_{N-1} \right]$$

Now make the following correspondences:

$m_1 \sim -m_1, \quad m_2 \sim -m_2, \quad \cdots \quad m_N \sim x, \quad$ and $\quad G's \sim Y's$.

The error terms are seen to be identical in form.

4.17 Transforming Eqs. (4.7.2) yields

$$G_1 S_{s+m_1} + G_2 S_s = S_s + A_1$$
$$G_1 S_s + G_2 S_{s+m_2} = S_s + A_2$$

where:
$$S_s = \frac{2}{-s^2+1}$$
$$S_{s+m_1} = \frac{-6s^2+12}{(-s^2+1)(-s^2+4)}$$
$$S_{s+m_2} = \frac{-s^2+3}{-s^2+1}$$

Solve for G_1 an G_2 to get the possible poles; e.g.

$$G_1 = A_1 \frac{-s^2+3}{\frac{-6s^2+20}{-s^2+4}} + A_2 \frac{2}{\frac{-6s^2+20}{-s^2+4}} + \frac{2}{\frac{-6s^2+20}{-s^2+4}}$$

Clearly, the only possible pole in the left half plane is at $-\sqrt{10}/3$. Therefore, following the hints in the problem statement, we can write G_1 and G_2 as

$$G_1 = K_1 + \frac{K_2}{s + \sqrt{10}/3}$$

$$G_2 = \frac{K_3}{s + \sqrt{10}/3}$$

K_1, K_2, K_3 are to be determined.

Next, substitute G_1 and G_2 into transformed equations above. After substitution, each term should be expanded into partial fraction form. Collect the coefficients of the $\frac{1}{s+1}$, $\frac{1}{s+2}$, and $\frac{1}{s+\sqrt{10}/3}$ terms and note that left and right sides of both equations must be identical in these terms. Three linear equations in $K_1 K_2$, and K_3 are then obtained and solved. The form of solution will depend on how the above procedure is carried out, and a variety of equivalent forms may be obtained. The easiest way to verify your solution is to look at the decimal equivalents: $K_1 \approx .4036774$, $K_2 \approx .070344$, $K_3 \approx .422064$

CHAPTER 5

5.1

First, perform spectral factorization.

$$S = \frac{-s^2 + 1}{s^4 - 8s^2 + 16} = \frac{s+1}{(s+2)^2} \cdot \frac{(-s+1)}{(-s+2)^2}$$

Following the modeling procedure of Sec 5.2 yields

Diff eq. for intermediate variable $r(t)$ is

$$\ddot{r} + 4\dot{r} + 4r = w(t)$$

Now define state variables as the phase variables:

$$x_1 = r$$
$$x_2 = \dot{r}$$

State equations are then

$$\begin{bmatrix} \dot{x}_1 \\ \dot{x}_2 \end{bmatrix} = \begin{bmatrix} 0 & 1 \\ -4 & -4 \end{bmatrix} \begin{bmatrix} x_1 \\ x_2 \end{bmatrix} + \begin{bmatrix} 0 \\ w(t) \end{bmatrix}$$

From the block diagram, 1 apparent that the scalar process y is the sum of x_1 and x_2. The measurement is of y directly (with additive noise), so the measurement model is

$$z_k = \begin{bmatrix} 1 & 1 \end{bmatrix} \begin{bmatrix} x_1 \\ x_2 \end{bmatrix} + \begin{bmatrix} \text{uncorrelated} \\ \text{noise} \end{bmatrix}$$

Therefore, H_k and R_k are obviously

$$H_k = \begin{bmatrix} 1 & 1 \end{bmatrix}$$

$$R_k = \begin{bmatrix} 2 \end{bmatrix} \quad \text{(from the problem statement)}$$

Transition matrix is obtained from the system F matrix.

$$\phi_R = \mathcal{L}^{-1}\left[(sI-F)^{-1}\right]_{t=1} = \mathcal{L}^{-1}\left[\begin{array}{cc} s & -1 \\ 4 & s+4 \end{array}\right]^{-1} \text{ Eval. at } t=1$$

$$= \left[\begin{array}{cc} e^{-2t}+2te^{-2t} & te^{-2t} \\ -4te^{-2t} & e^{-2t}-2te^{-2t} \end{array}\right]_{t=1} = \left[\begin{array}{cc} 3e^{-2} & e^{-2} \\ -4e^{-2} & -e^{-2} \end{array}\right]$$

The Q_R matrix is obtained as follows:

First, find transfer functions between input and the x_1 and x_2 responses. They are:

$$G_1 = \frac{1}{s^2+4s+4} \quad , \quad (w(t) \text{ to } x_1(t))$$

$$G_2 = \frac{s}{s^2+4s+4} \quad , \quad (w(t) \text{ to } x_2(t))$$

Now, we need $E(x_1^2(\Delta t))$, $E(x_2^2(\Delta t))$, and $E[x_1(\Delta t) x_2(\Delta t)]$. They are obtained from the nonstationary methods given in chap. 3. Note the weighting functions are:

$$g_1(u) = \mathcal{L}^{-1}[G_1(s)] = te^{-2t}$$

$$g_2(u) = \mathcal{L}^{-1}[G_2(s)] = e^{-2t}-2te^{-2t}$$

The needed variances and covariances are then

$$E\{x_1^2(\Delta t)\} = \int_0^{\Delta t}\int_0^{\Delta t} g_1(u)g_1(v) R_w(u-v)\,du\,dv$$

$$= \int_0^{\Delta t}\int_0^{\Delta t} g_1(u)g_1(v) S(u-v)\,du\,dv = \int_0^1 g_1^2(v)\,dv$$

Similarly,

$$E[x_2^2(\Delta t)] = \int_0^1 g_2^2(v)\,dv \quad , \quad \text{and}$$

$$E[x_1(\Delta t)\cdot x_2(\Delta t)] = \int_0^1 g_1(v)g_2(v)\,dv$$

It is now a routine matter to evaluate the integrals. We will leave the result in the form:

$$Q_k = \begin{bmatrix} \int_0^t g_1^2(v)\,dv & \int_0^t g_1(v)g_2(v)\,dv \\ \int_0^t g_1(v)g_2(v)\,dv & \int_0^t g_2^2(v)\,dv \end{bmatrix}$$

The initial conditions:

Both x_1 (i.e., y) and x_2 (its derivative) are zero-mean processes, so the best initial (i.e., a priori) estimate of each is zero.

$$\therefore \hat{\underline{x}}^- = \begin{bmatrix} 0 \\ 0 \end{bmatrix}$$

The initial P calculation is more complicated. It is:

$$P_0^- = \begin{bmatrix} E[x_1^2(0)] & E[x_1(0)\cdot x_2(0)] \\ E[x_1(0)\cdot x_2(0)] & E[x_2^2(0)] \end{bmatrix}$$

The x_1 and x_2 processes are stationary, so stationary methods or transient methods with $t=\infty$ may be used. The latter leads to:

$$P_0^- = \begin{bmatrix} \text{Same expression as for } Q_k, \text{ but} \\ \text{with the upper limits replaced with } \infty. \end{bmatrix}$$

The "11" and "22" terms are just the mean square values of x_1 and x_2, and are also easily obtained using integral tables on p. 126. However, the "12" term is not so easily evaluated, so the "transient" formula, namely $\int_0^t g_1(v)g_2(v)\,dv$, is probably the simplest approach.

Let $x_1 = s(t)$, $x_2 = m(t)$

Shaping filters are then:

$w_s(t) \longrightarrow \boxed{\dfrac{\sqrt{2\sigma_s^2 \beta_s}}{s + \beta_s}} \longrightarrow x_1$; $w_m(t) \longrightarrow \boxed{\dfrac{\sqrt{2\sigma_m^2 \beta_m}}{s + \beta_m}} \longrightarrow x_2$

(unity white noise)   (unity white noise)

Differential Eqs.:

$$\dot{x}_1 + \beta_s x_1 = \sqrt{2\sigma_s^2 \beta_s}\; w_s(t)$$

$$\dot{x}_2 + \beta_m x_2 = \sqrt{2\sigma_m^2 \beta_m}\; w_m(t)$$

State model:

$$\begin{bmatrix} \dot{x}_1 \\ \dot{x}_2 \end{bmatrix} = \begin{bmatrix} -\beta_s & 0 \\ 0 & -\beta_m \end{bmatrix} \begin{bmatrix} x_1 \\ x_2 \end{bmatrix} + \begin{bmatrix} \sqrt{2\sigma_s^2 \beta_s}\; w_s(t) \\ \sqrt{2\sigma_m^2 \beta_m}\; w_m(t) \end{bmatrix}$$

State variables are decoupled, so ϕ_k is obtained by inspection as:

$$\phi_k = \begin{bmatrix} e^{-\beta_s \Delta t} & 0 \\ 0 & e^{-\beta_m \Delta t} \end{bmatrix}$$

Measurement is additive combination of x_1 and x_2

Thus $\quad z = x_1 + x_2$

or $\quad z = \begin{bmatrix} 1 & 1 \end{bmatrix} \begin{bmatrix} x_1 \\ x_2 \end{bmatrix} + [0]$

$\therefore\ H_k = \begin{bmatrix} 1 & 1 \end{bmatrix}$; $R_k = [0]$ (Permissible in the discrete model)

Driven responses are obtained by methods of Chap. 3.

For example:

$$E[x_1^2(\Delta t)] = \int_0^{\Delta t} \int_0^{\Delta t} \sqrt{2\sigma_s^2 \beta_s}\; e^{-\beta_s u}\; \sqrt{2\sigma_s^2 \beta_s}\; e^{-\beta_s v} \cdot \delta(u-v)\, du\, dv$$

$$= 2\sigma_s^2 \beta_s \int_0^{\Delta t} e^{-2\beta_s v}\, dv = \sigma_s^2 (1 - e^{-2\beta_s \Delta t})$$

Therefore,

$$Q_k = \begin{bmatrix} \sigma_s^2(1 - e^{-2\beta_s \Delta t}) & 0 \\ 0 & \sigma_m^2(1 - e^{-2\beta_m \Delta t}) \end{bmatrix}$$

Initial Cond.:

$$\hat{x}_0^- = \begin{bmatrix} 0 \\ 0 \end{bmatrix} , \quad P_0^- = \begin{bmatrix} \sigma_s^2 & 0 \\ 0 & \sigma_m^2 \end{bmatrix}$$

5.3

$w(t) \rightarrow \boxed{\dfrac{2}{S}} \xrightarrow{x_2} \boxed{\dfrac{10}{S}} \xrightarrow{x_1}$

$x_1(0) = x_2(0) = 0$

$\mathbf{3}$ = Noisy measurement of x_1.

Diff. Eqs.:

$\dot{x}_1 = 10 x_2$

$\dot{x}_2 = 2 w(t)$

State Model:

$$\begin{bmatrix} \dot{x}_1 \\ \dot{x}_2 \end{bmatrix} = \begin{bmatrix} 0 & 10 \\ 0 & 0 \end{bmatrix} \begin{bmatrix} x_1 \\ x_2 \end{bmatrix} + \begin{bmatrix} 0 \\ 2w(t) \end{bmatrix}$$

Transition Matrix:

$$\Phi = \mathcal{L}^{-1}\{(sI - F)^{-1}\} = \mathcal{L}^{-1}\left\{\begin{bmatrix} s & -10 \\ 0 & s \end{bmatrix}^{-1}\right\} = \begin{bmatrix} 1 & 10t \\ 0 & 1 \end{bmatrix}$$

Evaluate for step size of 2 sec. This yields Φ_R.

$$\Phi_R = \begin{bmatrix} 1 & 20 \\ 0 & 1 \end{bmatrix}$$

The measurement is of x_1 directly, so

$$\mathbf{3} = \begin{bmatrix} 1 & 0 \end{bmatrix}\begin{bmatrix} x_1 \\ x_2 \end{bmatrix} + \begin{Bmatrix} \text{Uncorr.} \\ \text{Noise} \end{Bmatrix}$$

$\therefore$ $H_R = \begin{bmatrix} 1 & 0 \end{bmatrix}$, and $R_R = [4]$ $\quad$ (From problem statement)

Driven response calculation:

Trans. fcn. from w to x_1 = $G_1 = \dfrac{20}{S^2}$; $\therefore g_1(u) = 20u$

Trans. fcn. from w to x_2 = $G_2 = \dfrac{2}{S}$; $\therefore g_2(u) = 2$

$$E[x_1^2(\Delta t)] = \int_0^2 \int_0^2 20u \cdot 20v \cdot \delta(u-v)\, du\, dv = 3200/3$$

$$E\{x_2^2(\Delta t)\} = \int_0^2 \int_0^2 2 \cdot 2 \cdot \delta(u-v)\, du\, dv = 8$$

$$E\{x_1(\Delta t)x_2 \Delta t\} = \int_0^2 \int_0^2 20u \cdot 2 \cdot \delta(u-v)\, du\, dv = 80$$

$$\therefore Q_R = \begin{bmatrix} 3200/3 & 80 \\ 80 & 8 \end{bmatrix}$$

Initial Conditions:

$$\hat{x}_0^- = \begin{bmatrix} 0 \\ 0 \end{bmatrix}, \quad P_0^- = \begin{bmatrix} 0 & 0 \\ 0 & 0 \end{bmatrix} \quad \text{(States } x_1 \text{ and } x_2 \text{ are known to be zero at } t = 0.\text{)}$$

5.4

w_1 and w_2 are white.

$S_1 = 4 \ (ft/sec^2)^2/\frac{rad}{sec}$

$S_2 = 16 \ (ft/sec)^2/\frac{rad}{sec}$

Only the Q_K matrix is requested in this problem.

Find transfer functions first.

$$G(w_1 \to x_1) = G_{11} = \frac{1}{s^2} \ ; \ g_{11}(u) = u$$

$$G(w_1 \to x_2) = G_{12} = \frac{1}{s} \ ; \ g_{12}(u) = 1$$

$$G(w_2 \to x_1) = G_{21} = \frac{1}{s} \ ; \ g_{21}(u) = 1$$

$$G(w_2 \to x_2) = G_{22} = 0 \ ; \ g_{22}(u) = 0$$

Now note w_1 and w_2 are independent and find mean square responses.

$$E[x_1^2(\Delta t)] = \int_0^1 \int_0^1 u \cdot v \cdot 4 \cdot S(u-v) \, du \, dv$$
$$+ \int_0^1 \int_0^1 1 \cdot 1 \cdot 16 \cdot S(u-v) \, du \, dv$$
$$= \frac{4}{3} + 16 = \frac{52}{3} \ (ft)^2$$

$$E[x_2^2(\Delta t)] = \int_0^1 \int_0^1 1 \cdot 1 \cdot 4 \cdot S(u-v) \, du \, dv = 4 \ (ft/sec)^2$$

$$E[x_1(\Delta t)x_2(\Delta t)] = E\left[\int_0^{\Delta t} g_{11}(u) \, w_1(\Delta t - u) \, du \int_0^{\Delta t} g_{12}(v) \, w_1(\Delta t - v) \, dv\right]$$
$$= \int_0^1 \int_0^1 u \cdot 1 \cdot 4 \, S(u-v) \, du \, dv = 2 \ ft^2/sec$$

$$\therefore \quad Q_K = \begin{bmatrix} 52/3 & 2 \\ 2 & 4 \end{bmatrix}$$

(a) If the initial condition is a random variable $N(0,1)$, the initial P matrix must be changed to:

$$P_0^- = 1$$

Otherwise, the model parameters are unchanged.

(b) At $t=0$:

$$K_0 = P_0^- H^T (H P_0^- H^T + R)^{-1} = 1 \cdot 1 (1 \cdot 1 \cdot 1 + \tfrac{1}{4})^{-1} = \tfrac{4}{5}$$

$$\hat{x}_0^+ = \hat{x}_0^- + \tfrac{4}{5}(3_0 - 1 \cdot \hat{x}_0^-) = \tfrac{4}{5} 3_0$$

$$P_0 = (I - K_0 H_0) P_0^- = (1 - \tfrac{4}{5}) = \tfrac{1}{5}$$

Next, project ahead.

$$\hat{x}_1^- = \phi \hat{x}_0 = 1 \cdot (\tfrac{4}{5} 3_0)$$

$$P_1^- = \phi P_0 \phi^T + Q = 1 \cdot \tfrac{1}{5} \cdot 1 + 1 = \tfrac{6}{5}$$

At $t=1$:

$$K_1 = P_1^- H^T (H P_1^- H^T + R)^{-1} = \tfrac{6}{5}(\tfrac{6}{5} + \tfrac{1}{4})^{-1} = \tfrac{24}{29}$$

This gain is slightly more than the gain of Example 1. This is to be expected. There is more uncertainty in the a priori estimate and thus the filter gives the measurement more weight than in Example 1.

(c) The gain in this problem and that of Example 1 should approach the same value as the number of steps becomes large. The system is observable and driven by noise, so the filter should eventually "forget" the initial condition.

5.6

(a) Cycling through the recursive steps is routine, so only the results will be given.

After assimilating the measurement z_0:

$$\hat{x}_0 = \tfrac{1}{2} z_0$$

$$P_0 = \tfrac{1}{2}$$

After assimilating the measurement z_1:

$$\hat{x}_1 = \phi \hat{x}_0 + \left(\frac{1 - \tfrac{1}{2} e^{-.04}}{2 - \tfrac{1}{2} e^{-.04}} \right) [z_1 - \phi \hat{x}_0]$$

$\hat{x}_0$ may be substituted out using the results of the first step. The expression for $\hat{x}_1$ is then

$$\hat{x}_1 = \left(\frac{e^{-.02}}{4 - e^{-.04}} \right) z_0 + \left(\frac{2 - e^{-.04}}{4 - e^{-.04}} \right) z_1 \qquad (1)$$

(b) The weighting function (Wiener) approach:
Use Eq. (4.8.4) with an appropriate change in subscripts.

$$\begin{bmatrix} E(z_0^2) & E(z_0 z_1) \\ E(z_0 z_1) & E(z_1^2) \end{bmatrix} \begin{bmatrix} k_0 \\ k_1 \end{bmatrix} = \begin{bmatrix} E(z_0 x_1) \\ E(z_1 x_1) \end{bmatrix} \qquad (2)$$

The indicated expectations are:

$$E(z_0^2) = E\left[(x_0 + v_0)^2 \right] = E\left[x_0^2 + 2 x_0 v_0 + v_0^2 \right]$$
$$= E(x_0^2) + E(2 x_0 v_0) + E(v_0^2) = 1 + 1 = 2$$

Similarly,

$$E(z_1^2) = 2; \quad E(z_0 z_1) = E\left[(x_0 + v_0)(x_1 + v_1) \right] = E(x_0 x_1) = e^{-.02}$$

$$E(z_0 x_1) = E\left[(x_0 + v_0) x_1 \right] = e^{-.02}; \quad E(z_1 x_1) = E\left[(x_1 + v_1)(x_1) \right] = 1$$

Now, solving Eqs. (2) above yields

$$R_0 = \frac{e^{-.02}}{4 - e^{-.04}}; \quad k_1 = \frac{2 - e^{-.04}}{4 - e^{-.04}} \qquad \begin{array}{l} \text{(same as in} \\ \text{Eq. (1) above)} \end{array}$$

5.7

$\ddot{x} + \omega_r^2 x = 0$; Gen Sol.: $x(t) = x(0)\cos\omega_r t + \frac{\dot{x}(0)}{\omega_r}\sin\omega_r t$

(a) The model using phase variables:

Let $x_1 = x$, $x_2 = \dot{x}$

Then,

$$\begin{bmatrix} \dot{x}_1 \\ \dot{x}_2 \end{bmatrix} = \begin{bmatrix} 0 & 1 \\ -\omega_r^2 & 0 \end{bmatrix} \begin{bmatrix} x_1 \\ x_2 \end{bmatrix} + \begin{bmatrix} 0 \\ 0 \end{bmatrix}$$

$$\Phi = \mathcal{L}^{-1}\left[(sI - F)^{-1}\right] = \mathcal{L}^{-1}\begin{bmatrix} s & -1 \\ \omega_r^2 & s \end{bmatrix}^{-1}$$

Evaluating inverses and letting $t = \Delta t$ yields

$$\Phi_k = \begin{bmatrix} \cos\omega_r\Delta t & \frac{1}{\omega_r}\sin\omega_r\Delta t \\ -\omega_r\sin\omega_r\Delta t & \cos\omega_r\Delta t \end{bmatrix}$$

Assume we measure x (i.e., x_1) directly. Then

$$z = \begin{bmatrix} 1 & 0 \end{bmatrix}\begin{bmatrix} x_1 \\ x_2 \end{bmatrix} + \begin{bmatrix} \text{Uncorr.} \\ \text{Noise} \end{bmatrix}$$

and

$$H_k = \begin{bmatrix} 1 & 0 \end{bmatrix} \; ; \quad R_k = [R] \quad \text{(a scalar)}$$

The process is undriven, so

$$Q_k = \begin{bmatrix} 0 & 0 \\ 0 & 0 \end{bmatrix}$$

Initial Conditions:

$$\hat{x}_0^- = \begin{bmatrix} 0 \\ 0 \end{bmatrix} \; ; \quad P_0^- = \begin{bmatrix} \infty & 0 \\ 0 & \infty \end{bmatrix} \quad \begin{array}{l}\text{(Very large elements} \\ \text{on major diagonal.)}\end{array}$$

(b) The process is deterministic, so specification of the constants $x(0)$ and $\dot{x}(0)$ specifies $x(t)$ for $t > 0$. Therefore, try letting

$$x_1' = x(0) \;,\quad x_2' = \dot{x}(0) \quad \begin{bmatrix}\text{Use "prime" to} \\ \text{distinguish from} \\ \text{previous } x_1 \text{ and } x_2.\end{bmatrix}$$

State model:

$$\begin{bmatrix} \dot{x}_1' \\ \dot{x}_2' \end{bmatrix} = \begin{bmatrix} 0 & 0 \\ 0 & 0 \end{bmatrix}\begin{bmatrix} x_1' \\ x_2' \end{bmatrix} + \begin{bmatrix} 0 \\ 0 \end{bmatrix}$$

The following Kalman filter parameters are now obvious:

$$\Phi_k = 1, \quad Q_k = 0, \quad R_k = [R], \quad \hat{x}_0^- = \begin{bmatrix} 0 \\ 0 \end{bmatrix}, \quad P_0^- = \begin{bmatrix} \infty & 0 \\ 0 & \infty \end{bmatrix}$$

The measurement relationship is the same as before in terms of x. Therefore, we have

$$\mathfrak{z}_k = \begin{bmatrix} \cos w_r t_k & \frac{1}{w_r} \sin w_r t_k \end{bmatrix} \begin{bmatrix} x_i \\ \dot{x}_i \end{bmatrix} + [0]$$

or

$$H_k = \begin{bmatrix} \cos w_r t_k & \frac{1}{w_r} \sin w_r t_k \end{bmatrix}$$

Note that the measurement matrix varies with each step; i.e. $t_k = k\Delta t$, $k = 0, 1, 2 \cdots$ This is permissible

(C) The appropriate linear transformation is obtained from the general solution (given in part(a)) and its derivative. The result is:

$$\begin{bmatrix} x_1 \\ x_2 \end{bmatrix} = \begin{bmatrix} \cos w_r t & \frac{1}{w_r} \sin w_r t \\ -w_r \sin w_r t & \cos w_r t \end{bmatrix} \begin{bmatrix} x_1' \\ x_2' \end{bmatrix}$$

(This transformation is time varying, which is perfectly permissible, but it should not be singular. Otherwise, we could not go back and forth between the two models in either direction. Check on singular condition:

$$\text{Det} [\ ] = \cos^2 w_r t + \sin^2 w_r t = 1$$

The transformation is not singular.)

We already have the $\phi_R, H_R \cdots$ etc parameters worked out in Prob. 5.2, so there is no need to go back to the diff. eq. Begin with

$$x_{R+1} = \phi x_R + w_R$$

But

$$x_R' = A x_R \quad \text{or} \quad x_R = A^{-1} x_R' \; ; \; A = \begin{bmatrix} 1 & 1 \\ 0 & 1 \end{bmatrix}$$

Substituting for x_R yields $\qquad A^{-1} = \begin{bmatrix} 1 & -1 \\ 0 & 1 \end{bmatrix}$

$$A^{-1} x_{R+1}' = \phi A^{-1} x_R' + w_R$$

Or

$$x_{R+1}' = (A \phi A^{-1}) x_R' + (A w_R)$$

The new transition matrix is:

$$\phi' = A \phi A^{-1} = \begin{bmatrix} 1 & 1 \\ 0 & 1 \end{bmatrix} \begin{bmatrix} e^{-\beta_s \Delta t} & 0 \\ 0 & e^{-\beta_m \Delta t} \end{bmatrix} \begin{bmatrix} 1 & -1 \\ 0 & 1 \end{bmatrix}$$

$$= \begin{bmatrix} e^{-\beta_s \Delta t} & (-e^{-\beta_s \Delta t} + e^{-\beta_m \Delta t}) \\ 0 & e^{-\beta_m \Delta t} \end{bmatrix}$$

The new Q matrix is:

$$Q_R' = E[(A w_R)(A w_R)^T] = A Q_R A^T = \begin{bmatrix} 1 & 1 \\ 0 & 1 \end{bmatrix} \begin{bmatrix} q_{11} & 0 \\ 0 & q_{22} \end{bmatrix} \begin{bmatrix} 1 & 0 \\ 1 & 1 \end{bmatrix}$$

$$= \begin{bmatrix} (q_{11} + q_{22}) & q_{22} \\ q_{22} & q_{22} \end{bmatrix}, \quad \text{where} \quad \begin{aligned} q_{11} &= \sigma_s^2(1 - e^{-2\beta_s \Delta t}) \\ q_{22} &= \sigma_m^2(1 - e^{-2\beta_m \Delta t}) \end{aligned}$$

The measurement relationship is:

$$3_R = H_R x_R + v_R$$
$$= (H_R A^{-1}) x_R' + v_R$$

∴ New H and R are:

$$H_R' = H_R A^{-1} = \begin{bmatrix} 1 & 1 \end{bmatrix} \begin{bmatrix} 1 & -1 \\ 0 & 1 \end{bmatrix} = \begin{bmatrix} 1 & 0 \end{bmatrix}$$
$$R_R' = E(v_R^2) = 0 \quad (v_R \text{ is zero})$$

5.8 (cont.)

Initial Conditions:

$$\hat{x}_0'^- = \begin{bmatrix} 0 \\ 0 \end{bmatrix} \; ; \; P_0'^- = A P_0^- A^T = \begin{bmatrix} (\sigma_s^2 + \sigma_m^2) & \sigma_m^2 \\ \sigma_m^2 & \sigma_m^2 \end{bmatrix}$$

5.9

Suppose analyst A solves his estimation problem in the x-domain. He begins with an initial $\hat{x}_{k-1}$ and associated P_{k-1}. He then projects ahead and computes $\hat{x}_k$ according to

$$\hat{x}_k = \hat{x}_k^- + K_k (\mathfrak{z}_k - H_k \hat{x}_k^-)$$

OR

$$\hat{x}_k = \phi_{k-1} \hat{x}_{k-1} + K_k (\mathfrak{z}_k - H_k \phi_{k-1} \hat{x}_{k-1}) \qquad (1)$$

Next, suppose analyst B chooses to work the same problem in the x'-domain. He begins with an initial $\hat{x}_{k-1}'$ and P_{k-1}' that are related to A's values via

$$\hat{x}_{k-1}' = A \hat{x}_{k-1} \quad , \quad and \quad P_{k-1}' = A P_{k-1} A^T$$

Analyst B then projects ahead and computes his estimate according to

$$\hat{x}_k' = \phi_{k-1}' \hat{x}_{k-1}' + K_k' (\mathfrak{z}_k - H_k' \phi_{k-1}' \hat{x}_{k-1}') \qquad (2)$$

We wish to show equivalence. We will use the transformation results of Prob 5.8 freely and begin with the K_k' term

$$K_k' = P_k'^- H_k'^T (H_k' P_k'^- H_k'^T + R_k)^{-1}$$

$$= (A P_k^- A^T)(A^{-1})'H_k^T [H_k A^{-1}(A P_k^- A^T)(A^{-1})'H_k^T + R_k]^{-1}$$

94

Cancelling terms where appropriate leads to:

$$K_R' = A\left[P_R^- H_R^T (H_R P_R^- H_R^T + R_R)^{-1}\right] = A K_R$$

Now, substituting out all "primed" terms in Eq (2) yields

$$\hat{\chi}_k' = (A\phi_{k-1} A')(A\hat{\chi}_{k-1}) + A K_k \left[\mathfrak{Z}_k - (H_R A^{-1})(A\phi_{k-1} A^{-1}) A \hat{\chi}_{k-1}\right]$$

$$= A\left[\phi_{k-1}\hat{\chi}_{k-1} + K_k (\mathfrak{Z}_k - H_R \phi_{k-1}\hat{\chi}_{k-1})\right] \qquad (3)$$

Finally, compare this result with Eq. (1), and it is clear that B's estimate is just A's estimate transformed via the A matrix.

<u>5.10</u>

$$P = P^- - KHP^- - P^- H^T K^T + K(HP^- H^T + R)K^T$$

First, note $P^- H^T K^T = (KHP^-)^T$, so they each have the same trace. Next, use the differentiation formulas given, and differentiate trace P with respect to K and set equal to zero.

$$\frac{d(\text{trace } P)}{dK} = -(HP^-)^T - (HP^-)^T + 2K(HP^- H^T + R) = 0$$

Now solve for K:

$$K = P^- H^T (HP^- H^T + R)^{-1}$$

(<u>Note</u>: The above derivation looks easy. However, there are 2 minor disadvantages:

    (1) One must know the 2 matrix differentiation formulas, and

    (2) The diff. calculus approach only says that the above K yields an extremum. We also need to show that tr P is a minimum.

<u>5.11</u>

$$f_{x^*|3} = \frac{f_{3|x^*} \, f_{x^*}}{f_3}$$

All densities are normal, so we can write

$$f_{3|x^*} = \frac{1}{(2\pi)^{\frac{n}{2}} |R|^{\frac{1}{2}}} e^{-\frac{1}{2}(3 - Hx^*)^T R^{-1}(3 - Hx^*)}$$

$$f_{x^*} = \frac{1}{(2\pi)^{\frac{n}{2}} |P^*|^{\frac{1}{2}}} e^{-\frac{1}{2}(x^* - m^*)^T P^{*-1}(x^* - m^*)}$$

$$f_3 = \frac{1}{(2\pi)^{\frac{n}{2}} \underbrace{|HP^*H^T + R|}_{C_3}^{\frac{1}{2}}} e^{-\frac{1}{2}(3 - m_3)^T C_3^{-1}(3 - m_3)}$$

$$\therefore \; f_{x^*|3} = \frac{|C_3|^{\frac{1}{2}}}{(2\pi)^{\frac{n}{2}} |R|^{\frac{1}{2}} |P^*|^{\frac{1}{2}}} e^{-\frac{1}{2}[\text{Sum of above exponents}]}$$

Writing out the indicate sum yields

$$[\;] = (3 - Hx^*)^T R^{-1}(3 - Hx^*) + (x^* - m^*)^T P^{*-1}(x^* - m^*)$$
$$- (3 - m_3)^T C_3^{-1}(3 - m_3)$$

Now, expand out the sum and collect quadratic, linear, and zero-order terms in x^*. This leads to:

$$[\;] = x^{*T}(H^T R^{-1} H + P^{*-1})x^*$$
$$- x^{*T}(H^T R^{-1} 3 + P^{*-1} m^*) - (3^T R^{-1} H + m^{*T} P^{*-1})x^*$$
$$+ [\text{zero-order terms}]$$

The desired form is:
$$(x^* - m)^T C^{-1}(x^* - m)$$

It is now obvious that
$$C^{-1} = H^T R^{-1} H + P^{*-1}$$

or
$$C = [P^{*-1} + H^T R^{-1} H]^{-1}$$

Comparison of the linear terms leads to:

$$P^{-1}m = H^T R^{-1} \underline{z} + P^{*-1} m^*$$

or

$$m = (PH^T R^{-1})\underline{z} + P P^{*-1} m^*$$

Now note that $PH^T R^{-1} = K$ and $P = (I - KH)P^*$
Substitution in equation for m yields

$$m = m^* + K(\underline{z} - Hm^*)$$

and K may be written as $P^* H^T (HP^* H^T + R)^{-1}$ to
obtain Eq (5.6.9). (Note: Interchanging alternative
forms for K and P has nothing to do with estimation
theory. The "legality" follows directly from certain
matrix identities that only require symmetry and
existence of inverses.)

unity
w.n.
$w(t)$

$i.c. = 0$

The process diff. eq.:

$$\dot{x} = w(t)$$

Solution:

$$x_{k+1} = 1 \cdot x_k + w_k \qquad (1)$$

where

$$E(w_k^2) = \int_0^{.1} \int_0^{.1} 1 \cdot 1 \cdot \delta(u-v)\, du\, dv = 0.1$$

$\therefore w_k \sim N(0, .1)$ (or std. Dev. of $w_k = \sqrt{.1}$)

We have available $u_0, u_1, u_2, \ldots$ that are $N(0,1)$.
Therefore, let x_0 in Eq. (1) above be zero and
generate x_k's according to:

$$x_{k+1} = x_k + (\sqrt{.1})u_k, \quad k = 0, 1, 2, \ldots \quad (\text{and } x_0 = 0)$$

5.13

A FORTRAN program for generating the desired Gauss-Markov sample sequence and the optimal filter is listed below. The program output is listed on the next page. The listing of the input set of random numbers $N(0,1)$ is omitted to save space. (This was done in the program just for checking purposes.)

```
C    PROBLEM 5.13   BROWN-- RANDOM SIGNALS
C    FOR CONVENIENCE USE SUBSCRIPTED VARIABLES
      REAL DUM(102),XTRUE(51),XHAT(51),XHATMN(52),PPLUS(51),
     Z PMINUS(52),GAIN(51),VK(51),WK(50),ZK(51),PHI
      INTEGER I,J,K
C    READ IN RANDOM NUMBERS N(0,1) UNFORMATED
      READ, DUM
C    LIST RANDOM NUMBERS UNFORMATED
      PRINT, DUM
C    GENERATE TRUE X PROCESS.   TIME INDEX SHIFTED 1 UNIT
      XTRUE(1)=DUM(102)
      PHI=EXP(-.02)
      DO 10 I=1,50
         WK(I)=.1980165*DUM(I)
         J=I+1
         XTRUE(J)=PHI*XTRUE(I)+WK(I)
   10 CONTINUE
C    GENERATE MEASUREMENT PROCESS
      DO 15 I=1,51
         J=I+50
         VK(I)=DUM(J)
         ZK(I)=XTRUE(I)+VK(I)
   15 CONTINUE
C    MAIN KALMAN FILTER LOOP
      XHATMN(1)=0.
      PMINUS(1)=1.
      DO 20 I=1,51
         GAIN(I)=PMINUS(I)/(PMINUS(I)+1.)
         XHAT(I)=XHATMN(I)+GAIN(I)*(ZK(I)-XHATMN(I))
         PPLUS(I)=(1.-GAIN(I))*PMINUS(I)
         J=I+1
         XHATMN(J)=PHI*XHAT(I)
         PMINUS(J)=PHI*PHI*PPLUS(I)+.03921
   20 CONTINUE
      WRITE (6,100)
  100 FORMAT (5X,'K',3X,'XTRUE',4X,'XHAT',3X,'PPLUS')
      DO 25 I=1,51
         K=I-1
C    LIST THE OUTPUT X, XHAT, AND P MATRIX
         WRITE(6,110) K,XTRUE(I),XHAT(I),PPLUS(I)
  110 FORMAT (4X,I3,3F8.3)
   25 CONTINUE
      STOP
      END
```

5.13 (cont.) The Ouput:

K	X TRUE	XHAT	PPLUS
0	-0.493	0.238	0.500
1	-0.821	0.348	0.342
2	-0.724	-0.109	0.269
3	-0.570	0.171	0.229
4	-0.742	-0.188	0.206
5	-0.551	-0.578	0.192
6	-0.580	-0.651	0.183
7	-0.371	-0.692	0.177
8	-0.203	-0.571	0.173
9	-0.516	-0.557	0.170
10	-0.552	-0.539	0.169
11	-0.538	-0.421	0.168
12	-0.570	-0.685	0.167
13	-0.582	-0.687	0.166
14	-0.596	-0.550	0.166
15	-0.391	-0.480	0.166
16	-0.005	-0.289	0.166
17	0.067	-0.178	0.165
18	0.028	0.033	0.165
19	0.260	0.053	0.165
20	0.137	0.173	0.165
21	0.237	0.170	0.165
22	0.202	0.285	0.165
23	0.237	0.268	0.165
24	0.258	0.037	0.165
25	0.088	-0.073	0.165
26	0.583	-0.228	0.165
27	0.549	0.029	0.165
28	0.537	0.125	0.165
29	0.456	0.152	0.165
30	0.256	0.048	0.165
31	0.204	-0.080	0.165
32	0.532	0.174	0.165
33	0.355	-0.028	0.165
34	0.541	-0.086	0.165
35	0.458	-0.087	0.165
36	0.317	-0.206	0.165
37	0.102	0.011	0.165
38	-0.098	0.048	0.165
39	-0.175	0.207	0.165
40	0.064	0.028	0.165
41	0.208	0.115	0.165
42	0.179	0.208	0.165
43	0.190	0.320	0.165
44	0.354	0.218	0.165
45	0.268	0.219	0.165
46	0.335	0.043	0.165
47	0.554	0.445	0.165
48	0.764	0.663	0.165
49	0.888	0.672	0.165
50	1.085	0.443	0.165

Sketch of process and estimates: Simulation appears to be typical.

99

CHAPTER 6

6.1 The state equations:

$$\begin{bmatrix} \dot{x}_1 \\ \dot{x}_2 \\ \dot{x}_3 \end{bmatrix} = \begin{bmatrix} 0 & \Omega_3 & 0 \\ -\Omega_3 & 0 & \Omega_x \\ 0 & -\Omega_x & 0 \end{bmatrix} \begin{bmatrix} x_1 \\ x_2 \\ x_3 \end{bmatrix} + \begin{bmatrix} 1 & 0 & 0 \\ 0 & 1 & 0 \\ 0 & 0 & 1 \end{bmatrix} \begin{bmatrix} \epsilon_x \\ \epsilon_y \\ \epsilon_3 \end{bmatrix}$$

Eigenvalues: Find roots of $|\lambda I - F|$.

$$|\lambda I - F| = \begin{vmatrix} \lambda & -\Omega_3 & 0 \\ \Omega_3 & \lambda & -\Omega_x \\ 0 & \Omega_x & \lambda \end{vmatrix} = \lambda\left[\lambda^2 + \Omega_x^2 + \Omega_3^2\right]$$

Note that $\Omega_x^2 + \Omega_3^2 = \Omega$ (Earth rate).

∴ Eigenvalues are $0, \pm j\Omega$

6.2 Define state variables as:

$$x_1 = \psi_x , \quad x_2 = \psi_y , \quad x_3 = \psi_3$$
$$x_4 = \epsilon_x , \quad x_5 = \epsilon_y , \quad x_6 = \epsilon_3$$

State equations are then:

$$\begin{bmatrix} \dot{x}_1 \\ \dot{x}_2 \\ \dot{x}_3 \\ \dot{x}_4 \\ \dot{x}_5 \\ \dot{x}_6 \end{bmatrix} = \left[\begin{array}{ccc|c} 0 & \Omega_3 & 0 & \\ -\Omega_3 & 0 & \Omega_x & I \\ 0 & -\Omega_x & 0 & \\ \hline & & & \\ & 0 & & 0 \\ & & & \end{array}\right] \begin{bmatrix} x_1 \\ x_2 \\ x_3 \\ x_4 \\ x_5 \\ x_6 \end{bmatrix} + \begin{bmatrix} 0 \end{bmatrix}$$

For the transition matrix, use the approx.:

$$\Phi = e^{F\Delta t} \approx I + F\Delta t + \frac{F^2 \Delta t^2}{2} + \cdots$$

Neglect 2$\underline{nd}$ order and higher terms in Δt. Then Φ_k is

$$\Phi_k \approx I + F\Delta t = \begin{bmatrix} 1 & \Omega_3 \Delta t & 0 & \Delta t & 0 & 0 \\ -\Omega_3 \Delta t & 1 & \Omega_1 \Delta t & 0 & \Delta t & 0 \\ 0 & -\Omega_1 \Delta t & 1 & 0 & 0 & \Delta t \\ & & & & & \\ & O & & & I & \\ & & & & & \end{bmatrix}$$

The system is undriven. Thus

$$Q_k = 0 \quad [\text{a } 6\times6 \text{ null matrix}]$$

We have measurements of x_1 and x_2. Thus

$$H_k = \begin{bmatrix} 1 & 0 & 0 & 0 & 0 & 0 \\ 0 & 1 & 0 & 0 & 0 & 0 \end{bmatrix}$$

Measurement errors are uncorrelated, so

$$R_k = \begin{bmatrix} \sigma_x^2 & 0 \\ 0 & \sigma_y^2 \end{bmatrix}$$

Initial Conditions:

$$\hat{x}_0^- = \begin{bmatrix} 0 \\ 0 \\ 0 \\ 0 \\ 0 \\ 0 \end{bmatrix}, \quad P_0^- = \begin{bmatrix} \sigma_\psi^2 & 0 & 0 & & & \\ 0 & \sigma_\psi^2 & 0 & & O & \\ 0 & 0 & \sigma_\psi^2 & & & \\ & & & \sigma_\varepsilon^2 & 0 & 0 \\ & O & & 0 & \sigma_\varepsilon^2 & 0 \\ & & & 0 & 0 & \sigma_\varepsilon^2 \end{bmatrix}$$

6.3

Form the M matrix indicated in problem statement. Note in this case only a 6×10 matrix must be investigated for rank rather than a 6×12. (See Chen reference cited at end of Chap. 4.)

Omitting the algebra, the columns of M are:

Col. # →

$$
\begin{array}{c}
 \\
M=
\end{array}
\begin{array}{cccccccc}
\textcircled{1} & \textcircled{2} & \textcircled{3} & \textcircled{4} & \textcircled{5} & \textcircled{6} & \textcircled{7} & \textcircled{8} \\
1 & 0 & 0 & -\Omega_3 & -\Omega_3^2 & 0 & 0 & \frac{\Omega_3}{3}\Omega^2 \\
0 & 1 & \frac{\Omega_3}{3} & 0 & 0 & -\Omega^2 & -\frac{\Omega_3}{3}\Omega^2 & 0 \\
0 & 0 & 0 & \Omega_x & \Omega_x\Omega_3 & 0 & 0 & -\Omega_x\Omega^2 \\
0 & 0 & 1 & 0 & 0 & -\Omega_3 & -\Omega_3^2 & 0 \\
0 & 0 & 0 & 1 & \frac{\Omega_3}{2} & 0 & 0 & -\Omega^2 \\
0 & 0 & 0 & 0 & 0 & \Omega_x & \Omega_x\frac{\Omega_3}{3} & 0
\end{array}
$$

$$
\begin{array}{cc}
\textcircled{9} & \textcircled{10} \\
\Omega_3^2\Omega^2 & 0 \\
0 & \Omega_3^2\Omega^2+\Omega_x^2\Omega^2 \\
-\Omega_x\frac{\Omega_3}{3}\Omega^2 & 0 \\
0 & \frac{\Omega_3}{3}\Omega^2 \\
-\frac{\Omega_3}{3}\Omega^2 & 0 \\
0 & -\Omega_x\Omega^2
\end{array}
$$

Usually, it is very difficult to check a general 6×10 matrix for rank. However, this one is exceptional and can be tested by investigating independence of column vectors.

First, look at cols. ①②③ and ④. These are ind. because the rank is 4. (Con't on next page.)

6.3 (cont.)

Next, include cols. ⑤ and ⑥ and note that col.⑤ is just [col.④]·Ω_3. Thus, throw out col.⑤ and consider cols. ⑦ and ⑧. Note

$$\text{col. ⑦} = [\text{col.⑥}]·\Omega$$

$$\text{col. ⑧} = [\text{col.④}]·(-\Omega^2)$$

Therefore, throw out cols. ⑦ and ⑧.

Finally, look at cols. ⑨ and ⑩ and note:

$$\text{col. ⑨} = [\text{col. 4}]·(-\Omega_3 \Omega^2)$$

$$\text{col. ⑩} = [\text{col. 6}]·(-\Omega^2)$$

Therefore, only ①,②,③,④, and ⑥ are independent and rank of M is thus 5. System is <u>not</u> observable.

6.4

$$\begin{bmatrix} 3'_1 \\ 3'_2 \end{bmatrix} = \begin{bmatrix} 1 & 0 \\ C_1 & C_2 \end{bmatrix} \begin{bmatrix} 3_1 \\ 3_2 \end{bmatrix}, \quad \text{or } \underline{3}' = C \underline{3}$$

"New" measurement relationship is

$$\underline{3}' = C[Hx + \sigma] = (CH)x + C\sigma$$

New H matrix is CH and meas. error is $C\sigma$. We require that $\text{Cov}(C\sigma)$ be diagonal.

$$\text{Cov}(C\sigma) = E\left\{ \begin{bmatrix} \sigma_1 \\ C_1\sigma_1 + C_2\sigma_2 \end{bmatrix} \begin{bmatrix} \sigma_1 & C_1\sigma_1 + C_2\sigma_2 \end{bmatrix} \right\}$$

$$= \begin{bmatrix} E(\sigma_1\sigma_1) & E[\sigma_1(C_1\sigma_1 + C_2\sigma_2)] \\ E[\sigma_1(C_1\sigma_1 + C_2\sigma_2)] & E[(C_1\sigma_1 + C_2\sigma_2)^2] \end{bmatrix}$$

<u>6.4</u> (con't.)

(a) Arbitrarily let $c_1 = 1$; then

$$E[v_1^2 + c_2 v_1 v_2] = 0$$

or

$$c_2 = -\frac{E(v_1^2)}{E(v_1 v_2)} = -\frac{r_{11}}{r_{12}} \quad (r_{11} \text{ and } r_{12} \text{ are elements in original } r \text{ matrix.})$$

(b) New R matrix:

$$R' = \begin{bmatrix} E(v_1^2) & 0 \\ 0 & E(v_1 + (-\frac{r_{11}}{r_{12}})v_2)^2 \end{bmatrix}$$

$$= \begin{bmatrix} r_{11} & 0 \\ 0 & -r_{11} + \frac{r_{11}^2 r_{22}}{r_{12}^2} \end{bmatrix}$$

<u>6.5</u>

This is an unusual model for a random process, but it appears to fit the physical situation quite well in this case.

6.6   Work the (b) part first.

$$P_{k+1}^- = \phi_k P_k \phi_k^T + Q_k$$

But
$$P_k = (I - K_k H_k) P_k^-$$

$$= P_k^- - K_k H_k P_k^-$$

$$= P_k^- - P_k^- H_k^T (H_k P_k^- H_k^T + R_k)^{-1} H_k P_k^-$$

$$\therefore \ P_{k+1}^- = \phi \left[P_k^- - P_k^- H_k^T (H_k P_k^- H_k^T + R_k)^{-1} H_k P_k^- \right] \phi_k^T$$

$$+ Q_k$$

Now, work the (a) part.

$$P_{k+1} = (I - K_{k+1} H_{k+1}) P_{k+1}^-$$

$$= P_{k+1}^- - P_{k+1}^- H_{k+1}^T (H_{k+1} P_{k+1}^- H_{k+1}^T + R_{k+1})^{-1} H_{k+1} P_{k+1}^- \quad (1)$$

But P_{k+1}^- can be written explicitly in terms
of P_k:
$$P_{k+1}^- = \phi_k P_k \phi_k^T + Q_k \quad (2)$$

Eq.(2) can now be substituted into Eq.(1)
and the desired equation is obtained.

(The point of this problem is this: One does
not need to cycle through the separate steps
shown in Fig. 5.7 just to find the error covariance.
Its recursive equation stands by itself. Break-
ing the computation into steps is a matter of convenience.)

6.7

Look at the nominal filter first. The parameters
are: $\Phi = 1$, $Q = 1$, $\hat{x}_0^- = 0$
$H = 1$, $R = .1$, $P_0^- = 1$
Results for the first 3 recursive steps are:

Time	Gain K	Error Cov. P
0	10/11	1/11 ≈ .09090
1	120/131	12/131 ≈ .091603
2	1430/1561	143/1561 ≈ .091607

System reaches apprex. steady-state in 3 steps,
and the std. deviation of the error ≈ .30267

Now, suppose the true Q is 1.1, but we use the
incorrectly computed gain sequence above. To
get actual estimation error variance we must
cycle the above sequence of gains through

$$P_R = (1-K_R)P_R^-(1-K_R) + K_R P_R^- K_R$$

(In the projection step, be careful to use
the true Q of 1.1 and not 1.0.)

Results for Q = 1.1 are:

time	Error Cov.
0	1/11 ≈ .09090
1	.092308
2	.092317 (std.dev. ≈ .30384)

If we now make the comparison on the basis of
standard deviation, we see that the error only
increases about .38% due to a 10% modeling
error in the Q parameter -- low sensitivity!!
If the analysis is repeated for Q = .9,
approximately the same results are obtained.

6.8

$$y_1 = at_1 + b + v_1$$
$$y_2 = at_2 + b + v_2$$
$$y_3 = at_3 + b + v_3$$
$$\text{etc.}$$

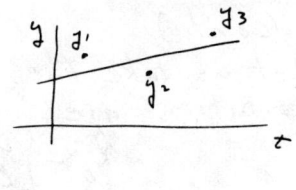

(a) One possible Kalman filter model:

Let a and b be parameters to be estimated. They are constants, so choose them as state variables with the state equation:

$$\begin{bmatrix} \dot{x}_1 \\ \dot{x}_2 \end{bmatrix} = \begin{bmatrix} 0 & 0 \\ 0 & 0 \end{bmatrix}\begin{bmatrix} x_1 \\ x_2 \end{bmatrix} + \begin{bmatrix} 0 \\ 0 \end{bmatrix} \; ; \quad \begin{array}{l} \text{where,} \\ x_1 = a \\ x_2 = b \end{array}$$

Discrete model is then:

$$\begin{bmatrix} x_1 \\ x_2 \end{bmatrix}_{k+1} = [I]\begin{bmatrix} x_1 \\ x_2 \end{bmatrix}_k + [0]$$

The process is undriven, so

$$Q_k = [0]$$

The measurements may be considered as a sequence of noisy scalars y_1, y_2, y_3 with an H matrix that changes with each step; e.g.,

$$[H]_{k=1} = [t_1 \quad 1]$$
$$[H]_{k=2} = [t_2 \quad 1]$$
$$[H]_{k=3} = [t_3 \quad 1]$$
$$\text{etc.}$$

Or, measurements may be batched and processed in one step. ("Time" is of no importance here.) For, say, 3 measurements H would be:

$$H = \begin{bmatrix} t_1 & 1 \\ t_2 & 1 \\ t_3 & 1 \end{bmatrix}$$

6.8 (cont.)

Let us take the "batched" viewpoint. The initial conditions are:

$$\hat{x}_o^- = \begin{bmatrix} 0 \\ 0 \end{bmatrix} \; ; \; P_o^- = \begin{bmatrix} \infty & 0 \\ 0 & \infty \end{bmatrix} \quad \text{(No prior knowledge of } x_1 \text{ and } x_2 \text{)}$$

Use the alternative algorithm of sec. 6.2, and process all k measurements in one step. Then

$$P^{-1} = (P^-)^{-1} + H^T R^{-1} H = (\infty)^{-1} + H^T R^{-1} H$$

$$= H^T R^{-1} H$$

Let the measurement covariance be:

$$R = \begin{bmatrix} r & 0 & 0 \\ 0 & r & 0 \\ 0 & 0 & r \ddots \end{bmatrix} = r I$$

Then

$$R^{-1} = \frac{1}{r} I$$

and P is

$$P = [H^T R^{-1} H]^{-1} = r (H^T H)^{-1}$$

The gain is then

$$K = P H^T R^{-1} = r (H^T H)^{-1} H^T \frac{1}{r}$$

Now, if no. of measurements $\geq$ no. of state variables to be estimated, $(H^T H)$ is invertable, and

$$\hat{x} = (H^T H)^{-1} H^T \underline{z} \; ; \; \underline{z} = \begin{bmatrix} y_1 \\ y_2 \\ \vdots \end{bmatrix}, \hat{x} = \begin{bmatrix} \hat{a} \\ \hat{b} \end{bmatrix}$$
$$\text{itc.}$$

(b) The demonstration requested can be done in terms of matrix components for the 3-tuple case. However, this algebra is really not needed. In Eq. (6.8.1) let $x = \begin{bmatrix} a \\ b \end{bmatrix}$, $M = H$, $b = 3$; and then in Eq. (6.8.5) let the weight factor W be I. It is then obvious that

$$\begin{bmatrix} a \\ b \end{bmatrix}_{opt.} = (H^T H)^{-1} H^T \underline{z}$$

108

6.9

z is measured from initial release point

m (Falling mass)

Ideal relationship:
$$\text{Distance} = \tfrac{1}{2} g t^2$$

Noisy measurement relationships:
$$z_1 = \left(\tfrac{1}{2} t_1^2\right) g + v_1$$
$$z_2 = \left(\tfrac{1}{2} t_2^2\right) g + v_2$$

etc.

(a) The Kalman filter model:

Let $x = g$ (a constant)

Then
$$\phi = 1$$
$$Q = 0$$

$$R = \begin{bmatrix} \sigma^2 & & & \\ & \sigma^2 & & \\ & & \sigma^2 & \\ & & & \ddots \end{bmatrix} = \sigma^2 I \quad (I \text{ is an } N \times N \text{ identity matrix})$$

$$H = \begin{bmatrix} \tfrac{1}{2} t_1^2 \\ \tfrac{1}{2} t_2 \\ \vdots \\ \tfrac{1}{2} t_N^2 \end{bmatrix},$$

(Note that we are considering all measurements batched into 1 vector in this model.)

$$\hat{x}_0^- = 0 \quad, \quad P_0^- = \infty \quad (\text{No prior knowledge})$$

This example is simple enough to be feasible to write out an explicit expression for $\hat{x}$ (or $\hat{g}$). Use of the alternative algorithm of Sec. 6.2 yields

$$\hat{x} = \frac{2 t_1^2}{\sum\limits_{i}^{N} t_i^4} z_1 + \frac{2 t_2^2}{\sum\limits_{i}^{N} t_i^4} z_2 + \cdots \cdot \frac{2 t_N^2}{\sum\limits_{i}^{N} t_i^4} z_N$$

6.9 (cont.)

(b) In the 3-state model let x be distance. Then
$$\ddot{x} = g$$
Choose state variables as
$$x_1 = x, \quad x_2 = \dot{x}, \quad x_3 = g$$
State model is then:

$$\begin{bmatrix} \dot{x}_1 \\ \dot{x}_2 \\ \dot{x}_3 \end{bmatrix} = \begin{bmatrix} 0 & 1 & 0 \\ 0 & 0 & 1 \\ 0 & 0 & 0 \end{bmatrix} \begin{bmatrix} x_1 \\ x_2 \\ x_3 \end{bmatrix} + [0]$$

Transition matrix for step size Δt is:
$$\Phi = \begin{bmatrix} 1 & \Delta t & \Delta t/2 \\ 0 & 1 & \Delta t \\ 0 & 0 & 1 \end{bmatrix}$$

Also,
$$H = [1 \ 0 \ 0], \quad Q = [0], \quad R = \sigma^2$$

(Note that measurements are not batched in this model.)

We begin at $t=0$ and use the direct algorithm of Chap. 5. The initial conditions are:

$$\hat{x}_0^- = [0], \quad P_0^- = \begin{bmatrix} 0 & 0 & 0 \\ 0 & 0 & 0 \\ 0 & 0 & \sigma_g^2 \end{bmatrix} \quad \begin{array}{l} \text{(Initial pos.} \\ \text{and vel. are} \\ \text{known to be zero.} \\ g \text{ is unknown.)} \end{array}$$

There is no measurement at $t=0$, so a posteriori $\hat{x}$ and P are same.

Next, project $\hat{x}_0$ and P_0 ahead to first measurement at $t = \Delta t$. This leads to:

$$P_1^- = \sigma_g^2 \begin{bmatrix} \Delta t^4/4 & \Delta t^3/3 & \Delta t^2/2 \\ \Delta t^3/3 & \Delta t^2 & \Delta t \\ \Delta t^2/2 & \Delta t & 1 \end{bmatrix}$$

and
$$\hat{x}_1^- = [0]$$

6.9 (cont.)

The gain is then found from $\bar{P}H^T(H\bar{P}H^T+R)^{-1}$.
The result is:

$$K_1 = \frac{\sigma_g^2}{\sigma_g^2 \Delta t^4/4 + \sigma^2} \begin{bmatrix} \Delta t^4/4 \\ \Delta t^3/2 \\ \Delta t^2/2 \end{bmatrix}$$

Now, let $\sigma_g^2 \to \infty$ indicating "infinite uncertainty" in knowledge of g initially.
This makes the gain approach

$$K_1 = \begin{bmatrix} 1 \\ 2/\Delta t \\ 2/\Delta t^2 \end{bmatrix} \qquad (\text{for } \sigma_g^2 = \infty)$$

The optimal estimate of g is then

$$\hat{g} = \frac{2}{\Delta t^2} \mathfrak{Z}_1$$

which checks with part (a).

Next, check on position and velocity estimates.
From Kalman filter estimates $\hat{x}_1$ and $\hat{x}_2$:

$$\hat{x}_1 = \hat{pos.} = 1 \cdot \mathfrak{Z}_1$$
$$\hat{x}_2 = \hat{vel.} = 2/\Delta t \cdot \mathfrak{Z}_1$$

If we were to use $\hat{g}$ and deterministic relationships:

$$\hat{pos.} = \tfrac{1}{2}\hat{g}\, t^2 = \tfrac{1}{2}\left(\tfrac{2}{\Delta t^2}\mathfrak{Z}_1\right)\Delta t^2 = \mathfrak{Z}_1 \quad \Big\}\ \substack{\text{same}\\ \text{as}}$$
$$\hat{vel.} = \hat{g}\, t = \left(\tfrac{2}{\Delta t^2}\mathfrak{Z}_1\right)(\Delta t) = 2/\Delta t \cdot \mathfrak{Z}_1 \quad \Big\}\ \substack{\text{K.F.}\\ \text{estimates}}$$

(The moral: states x_1 and x_2 are just excess baggage here!!)

///

CHAPTER 7

7.1

$R_w = 16\,\delta(t)$
$R_v = 4\,\delta(t)$

Filter parameters: $F=0$, $Q=16$, $R=4$, $H=1$, $G=1$

Ricatti Eq: $\dot{P} = -P^2/4 + 16$

or $\dfrac{dP}{-P^2 + 64} = dt/4$

The simplest solution is to expand left side with a partial fraction expansion. Then integrate term by term. This leads to

$$\frac{\frac{1}{16}\,dP}{P+8} + \frac{\frac{1}{16}\,dP}{-P+8} = \frac{dt}{4}$$

General Solution:

$$\frac{P+8}{P-8} = ce^{4t}, \quad \text{Boundary Cond.: } \begin{cases} P=0 \text{ when} \\ t=0 \end{cases}$$

C is determined from bound. cond. and is -1

$$\therefore \quad \frac{P+8}{P-8} = -e^{4t}, \quad \text{or} \quad P = 8\,\frac{1-e^{-4t}}{1+e^{-4t}}$$

(a) Optimal filter:

$$\dot{\hat{x}} = K(3-\hat{x}), \quad \begin{cases} K = P/4 \\ = 2\,\dfrac{1-e^{-4t}}{1+e^{-4t}} \end{cases}$$

(b) Steady-state:

$P=8$, $K=2$

$\therefore \quad \dot{\hat{x}} = 2(3-\hat{x})$, or

$\dot{\hat{x}} + 2\hat{x} = 2\,3$

or $\dfrac{\hat{x}(s)}{Z(s)} = \dfrac{2}{s+2}$ (Filter transfer function)

optimal steady-state filter:

Steady-state rms error
$= \sqrt{P} = \sqrt{8} = 2\sqrt{2}$

7.2 Begin with a state model defined in the usual way.

Let $x_1 = S(t)$,

$x_2 = m_1(t)$,

$$w_1(t) \rightarrow \boxed{\dfrac{\sqrt{2}}{S+1}} \rightarrow S(t) \quad (\text{or } x_1)$$

$$w_2(t) \rightarrow \boxed{\dfrac{2}{S+2}} \rightarrow m_1(t) \quad (\text{or } x_2)$$

w_1 and w_2 are ind unity white noise processes.

Then

$$\begin{Bmatrix} \dot{x}_1 \\ \dot{x}_2 \end{Bmatrix} = \begin{bmatrix} -1 & 0 \\ 0 & -2 \end{bmatrix} \begin{Bmatrix} x_1 \\ x_2 \end{Bmatrix} + \begin{bmatrix} \sqrt{2}\, w_1(t) \\ 2\, w_2(t) \end{bmatrix}$$

$$\begin{Bmatrix} 3_1 \\ 3_2 \end{Bmatrix} = \begin{bmatrix} 1 & 1 \\ 1 & 0 \end{bmatrix} \begin{Bmatrix} x_1 \\ x_2 \end{Bmatrix} + \begin{bmatrix} 0 \\ n_2(t) \end{bmatrix}$$

However, R matrix is singular for the above model, so make a transformation and eliminate one state variable from the estimation problem. Define a new set of state variables as:

$$\begin{Bmatrix} y_1 \\ y_2 \end{Bmatrix} = \begin{bmatrix} 1 & 1 \\ 0 & 1 \end{bmatrix} \begin{Bmatrix} x_1 \\ x_2 \end{Bmatrix}, \quad \text{or} \quad \underline{y} = \wedge \underline{x}$$

In the y-domain, the model becomes:

$$\begin{bmatrix} \dot{y}_1 \\ \dot{y}_2 \end{bmatrix} = \begin{bmatrix} -1 & -1 \\ 0 & -2 \end{bmatrix} \begin{Bmatrix} y_1 \\ y_2 \end{Bmatrix} + \begin{bmatrix} \sqrt{2}\, w_1 + 2 w_2 \\ 2 w_2 \end{bmatrix}$$

$$\begin{Bmatrix} 3_1 \\ 3_2 \end{Bmatrix} = \begin{bmatrix} 1 & 0 \\ 1 & -1 \end{bmatrix} \begin{Bmatrix} y_1 \\ y_2 \end{Bmatrix} + \begin{bmatrix} 0 \\ v_2 \end{bmatrix}, \quad v_2 = m_2$$

Now 3_1 is a perfect measurement of y_1, so it (ie., y_1) can be eliminated from estimation problem. The remaining state equation is then

$$\dot{y}_2 = -2 y_2 + 2 w_2(t)$$

This is in the correct form for a continuous K.F.

7.2 (cont.)

we must now find appropriate linear
connections between z_1 and z_2 and y_2.

Note $z_1 + \dot{z}_1 = y_1 + \dot{y}_1 = y_1 + (-y_1 - y_2 + \sqrt{2}\,w_1 + 2w_2) = -y_2 + \sqrt{2}\,w_1 + 2w_2$

and $z_1 - z_2 = y_2 - v_2$

Therefore, the meas. model for y_2 as the state variable
is:

$$\begin{bmatrix} z_1 - \dot{z}_1 \\ z_1 - z_2 \end{bmatrix} = \begin{bmatrix} -1 \\ 1 \end{bmatrix} [y_2] + \begin{bmatrix} \sqrt{2}\,w_1 + 2w_2 \\ -v_2 \end{bmatrix}$$

we can now find P from Eq (7.3.24). We only
want steady-state solution, so $\dot{P} = 0$ and problem
reduces to solving an algebraic equation in P.
The parameters are:

$F = -2$, $Q = 4$ (considering $2w_2$ as "input"), $G = 1$,

$R = \begin{bmatrix} 6 & 0 \\ 0 & 1 \end{bmatrix}$, $R^{-1} = \begin{bmatrix} 1/6 & 0 \\ 0 & 1 \end{bmatrix}$, $H = \begin{bmatrix} -1 \\ 1 \end{bmatrix}$,

C is obtained b noting

$E[\underline{w}\ \underline{v}^T] = E\left\{[2w_2][(\sqrt{2}\,w_1 + 2w_2)\ (-v_2)]\right\}$

$= [4 \quad 0]\, \delta(t - r)$

$\therefore C = [4 \quad 0]$, and thus

$D = GCR^{-1} = [1][4 \quad 0]\begin{bmatrix} 1/6 & 0 \\ 0 & 1 \end{bmatrix} = [2/3 \quad 0]$

Letting $\dot{P} = 0$ in Eq.(7.3.24) then leads to:

$-7/6\,P^2 - 8/3\,P + 4/3 = 0$

Choosing the positive root:

$P = \frac{1}{7}(-8 + 6\sqrt{10/3})$

114

7.2 (cont.)

The gain is obtained from Eq.(7.3.23):

$$K = (PH^T + GC)R^{-1} = \left[-\frac{P+4}{6} \quad P \right]$$

Finally, the equation for $\hat{y}_2$ becomes: (Eq. 7.3.22)

$$\dot{\hat{y}}_2 = (-2)\hat{y}_2 + \left[-\frac{P+4}{6} \quad P \right] \left[\begin{bmatrix} 3_1 - \dot{3}_1 \\ 3_1 - 3_2 \end{bmatrix} - \begin{bmatrix} -1 \\ 1 \end{bmatrix} \hat{y}_2 \right]$$

The above equation is the equivalent to:

3_2 ⟶ [$\frac{-P}{s+a}$] ⟶

3_1 ⟶ [$\frac{-\frac{P+4}{6}(s+1) + P}{s+a}$] ⟶ ⊕ ⟶ $\hat{y}_2$

$\left(a = \frac{8+7P}{6} \right)$

and

$P = \frac{1}{7}\left(-8 + 6\sqrt{\frac{10}{3}}\right)$

Also, $\hat{y}_1$ is given by the trivial block diagram:

3_1 ⟶ [1] ⟶ $\hat{y}_1$

Now, transform back to the x-domain via

$$\hat{x}_1 = \hat{y}_1 - \hat{y}_2$$
$$\hat{x}_2 = \hat{y}_2$$

(or equivalent in complex domain)

This will eventually lead to the transfer functions given in Problem 4.17. (This problem gives one an appreciation of the effort involved in solving even relatively simple multiple-input continuous estimation problems.)

7.3 Let $P_0 = 1$ in Eq (7.2.16). This yields

$$P(t) = \frac{\cosh \sqrt{3}t + 1/\sqrt{3} \cdot \sinh \sqrt{3}t}{\cosh \sqrt{3}t + 2/\sqrt{3} \cdot \sinh \sqrt{3}t}$$

At $t=0$, $P(0) = 1$; at $t = \infty$, $P(\infty) = \sqrt{3}-1 \approx .732$

Therefore, the general shape of $P(t)$ curve is as shown

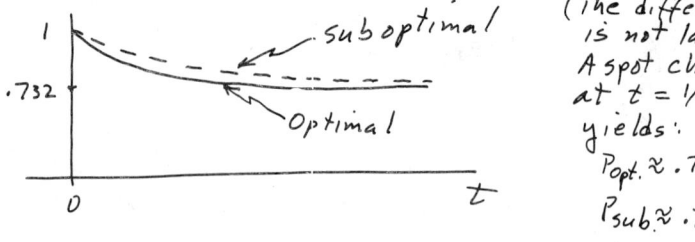

Suboptimal case:

Let $K = \sqrt{3} - 1$ in Eq (7.5.5). This yields

$$\dot{P} + 2\sqrt{3}\,P = 6 - 2\sqrt{3} \quad , \quad P(0) = 1$$

The solution has simple exponential form. As $t \to \infty$, solution should approach $\sqrt{3} - 1$. Check: Let $\dot{P} = 0$, then

$$P = \frac{6 - 2\sqrt{3}}{2\sqrt{3}} = \frac{3}{\sqrt{3}} - 1 = \sqrt{3} - 1$$

Comparison of optimal and suboptimal:

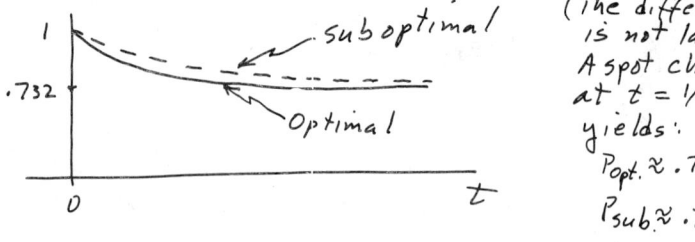

(The difference is not large. A spot check at $t = 1/\sqrt{3}$ yields:

$P_{opt.} \approx .76604$

$P_{sub.} \approx .76831$)

116

7.4

Parts (a) and (b) are very similar to Prob 5.6. Thus, only the results will be given. The optimal estimate is:

$$\hat{x}_1 = \frac{4e^{-1}}{25-16e^{-2}} \, \mathring{z}_0 + \left(\frac{20-16e^{-2}}{25-16e^{-2}}\right) \mathring{z}_1$$

(c) In this part we artificially let $P_0^- = \infty$. (In effect, this amounts to ignoring the prior knowledge of the variance of the process.) The "Kalman filter" solution for $P_0^- = \infty$ is:

$$\hat{x}_{\substack{sub-\\opt.}} = \left(\frac{e^{-1}}{5-3e^{-2}}\right) \mathring{z}_0 + \left(\frac{4-3e^{-2}}{5-3e^{-2}}\right) \mathring{z}_1$$

Clearly this is different from the results of (a) and (b). Strictly speaking this should <u>not</u> be called either a Wiener or Kalman estimate. <u>It is not</u> a <u>least squares estimate</u>!

7.5

This problem is similar to Prob. 7.7, but involves more algebra. It is workable though. Proceed as follows:

Use Eq (4.4.22) and write impulse response function.

$$h(t,r) = a(t)\ddot{e}^{-\sqrt{3}(t-r)} + b(t) e^{\sqrt{3}(t-r)}$$

Then $\hat{x}(t)$ may be written as

$$\hat{x}(t) = \int_0^t \left[a(t)\ddot{e}^{-\sqrt{3}t} e^{\sqrt{3}r} + b(t)e^{\sqrt{3}t}\ddot{e}^{-\sqrt{3}r}\right] \mathring{z}(r) \, dr$$

Now, show the above satifies:

$$\dot{\hat{x}} = F\hat{x} + K(z - H\hat{x})$$

117

First form left side of equation (ie., $\dot{\hat{x}}$)
This leads to:

$$\dot{\hat{x}} = [a+b]\, \underset{\sim}{3}(t) + \int_0^t \{[\dot{a} - \sqrt{3}a]\, \bar{e}^{\sqrt{3}t} e^{\sqrt{3}\tau}$$

$$+ [\dot{b} + \sqrt{3}b] e^{\sqrt{3}t} e^{-\sqrt{3}\tau}\}\, 3(\tau)\, d\tau$$

Next, note gain $K = P$, and thus right side
of diff. eq. is (noting $H = 1$ and $F = -1$):

$$(-1-P)\hat{x} + P\, 3(t)$$

Now, in order to have equality, linear terms in
$3(t)$ must be equal. Therefore show that

$$a + b = P$$

This is easily done with reference to Eqs.(7.2.16)
and (4.4.16) and (4.4.17) and some algebra. Next,
write $(-1-P)\hat{x}$ as

$$(-1-P)\hat{x} = (-1-P)\int_0^t [a(t)e^{-\sqrt{3}t} e^{\sqrt{3}\tau} + b(t)e^{\sqrt{3}t} e^{-\sqrt{3}\tau}]\, 3(\tau)\, d\tau$$

We now must show that

$$(-1-P)a(t) = \dot{a}(t) - \sqrt{3}\, a(t)$$

and

$$(-1-P)\, b(t) = \dot{b}(t) + \sqrt{3}\, b(t)$$

This involves some algebra and it is helpful
to show the equivalent relationships: (noting
that $P = a + b$)

$$\frac{\dot{a}}{a} + (-\sqrt{3}+1) = -(a+b)$$

$$\frac{\dot{b}}{b} + (\sqrt{3}+1) = -(a+b)$$

Once this is done, the sol. of the de. is verified.

7.6 $z(t) = a_0 + m(t)$, $t \geq 0$, $R_m(\tau) = A\delta(\tau)$

Let $a_0 = x$

The state model is then

$$\dot{x} = 0 \cdot x + 0$$
$$z = 1 \cdot x + v(t) \qquad [\text{Meas. eq., } v(t) = m(t)]$$

Filter parameters:

$H = 1$, $Q = 0$, $G = 1$, $F = 0$, $R = A$

Ricatti Eq:

$$\dot{P} = -\frac{1}{A} P^2$$

The variables are separable, so this is easily integrated. Using the boundary cond., $P(0) = \sigma^2$, leads to

$$P = \frac{1}{\frac{t}{A} + \frac{1}{\sigma^2}} = \frac{A\sigma^2}{\sigma^2 t + A}$$

The filter gain is then

$$K = P H^T R^{-1} = \frac{\sigma^2}{\sigma^2 t + A}$$

Filter block diagram:

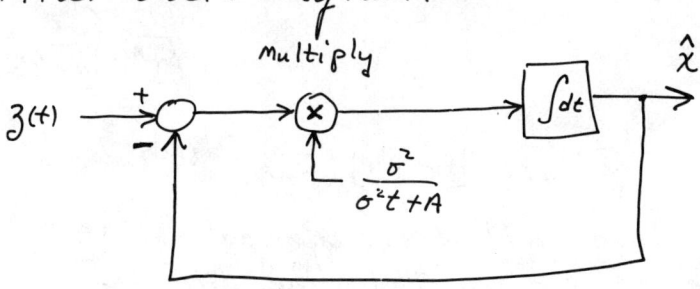

The Kal. filter diff. eq: $\dot{\hat{x}} = \left(\dfrac{1}{t + \frac{A}{\sigma^2}}\right)(z - \hat{x})$ (1)

Wiener solution:

$$\hat{x} = \int_0^t \frac{1}{\left(t + \frac{A}{\sigma^2}\right)} \cdot z(t-\gamma)\, d\gamma \quad (2)$$

To show equivalence, simply verify that the explicit Wiener solution will satisfy the Kalman filter diff. equation. First form derivative of $\hat{x}$ as per Eq (2):

$$\dot{\hat{x}} = \frac{d}{dt}\left[\frac{1}{t + \frac{A}{\sigma^2}}\int_0^t z(t-\gamma)\, d\gamma\right]$$

$$= \frac{1}{\left(t + \frac{A}{\sigma^2}\right)}\left[z(t-t) + \int_0^t \frac{\partial z(t-\gamma)}{\partial t}\, d\gamma\right]$$

$$+ \;-\frac{1}{\left(t + \frac{A}{\sigma^2}\right)^2}\int_0^t z(t-\gamma)\, d\gamma$$

$$= \frac{1}{t + \frac{A}{\sigma^2}} z(0) + \frac{1}{t + \frac{A}{\sigma^2}}\left[z(t) - z(0)\right]$$

$$- \frac{1}{t + \frac{A}{\sigma^2}}\int_0^t \frac{1}{t + \frac{A}{\sigma^2}} z(t-\gamma)\, d\gamma$$

$$= \frac{1}{t + \frac{A}{\sigma^2}}\left[z - \int_0^t \frac{1}{t + \frac{A}{\sigma^2}} z(t-\gamma)\, d\gamma\right]$$

The above will be recognized as the right side of Eq. (1). Thus the diff. eq. is satisfied and the Wiener and Kalman solutions are shown to be equivalent.

7.8

$$w \rightarrow \boxed{\frac{1}{5}} \xrightarrow{x_2} \boxed{\frac{1}{5}} \xrightarrow{x_1} \;\; \overset{+}{\underset{+}{\bigcirc}} \xleftarrow{v(t)} \rightarrow z$$

$R_w(\tau) = 10\, \delta(\tau)$

$R_v(\tau) = 100\, \delta(\tau)$

Process equations:

$$\dot{x}_1 = x_2 \quad , \quad \dot{x}_2 = w(t)$$

State and measurement model:

$$\begin{bmatrix} \dot{x}_1 \\ \dot{x}_2 \end{bmatrix} = \begin{bmatrix} 0 & 1 \\ 0 & 0 \end{bmatrix} \begin{bmatrix} x_1 \\ x_2 \end{bmatrix} + \begin{bmatrix} 0 \\ 1 \end{bmatrix} [w(t)]$$

$\Uparrow$ F matrix $\qquad$ $\Uparrow$ G matrix

$$Q = [10]$$
$$R = [100]$$
$$H = [1 \; 0]$$

Initial conditions:

$$\begin{bmatrix} \hat{x}_1 \\ \hat{x}_2 \end{bmatrix}_0^- = \begin{bmatrix} 0 \\ 0 \end{bmatrix} , \quad P_0^- = \begin{bmatrix} 1 & 0 \\ 0 & 0 \end{bmatrix}$$

Differential Eq. for P:

$$\dot{P} = FP + PF^T - PH^TR^{-1}HP + GQG^T, \quad P(0) = 0$$

or

$$\begin{bmatrix} \dot{P}_{11} & \dot{P}_{12} \\ \dot{P}_{12} & \dot{P}_{22} \end{bmatrix} = \begin{bmatrix} 0 & 1 \\ 0 & 0 \end{bmatrix} \begin{bmatrix} P_{11} & P_{12} \\ P_{12} & P_{22} \end{bmatrix} + \begin{bmatrix} P_{11} & P_{12} \\ P_{12} & P_{22} \end{bmatrix} \begin{bmatrix} 0 & 0 \\ 1 & 0 \end{bmatrix} - \begin{bmatrix} P_{11} & P_{12} \\ P_{12} & P_{22} \end{bmatrix} \begin{bmatrix} 1 \\ 0 \end{bmatrix} [100]$$

$$[1 \; 0] \begin{bmatrix} P_{11} & P_{12} \\ P_{12} & P_{22} \end{bmatrix} + \begin{bmatrix} 0 \\ 1 \end{bmatrix} [10][0 \; 1]$$

7.9

Begin with the discrete estimation equation:

$$\hat{x}_k = \hat{x}_{\bar{k}} + K_k(z_k - H_k\hat{x}_{\bar{k}}) \qquad (1)$$

We assume here that $\hat{x}_{\bar{k}}$ includes the effect of the deterministic input; i.e.,

$$\hat{x}_{\bar{k}} = \phi_{k-1}\hat{x}_{k-1} + \delta x \qquad \left(\begin{array}{l}\delta x \text{ is the} \\ \text{deterministic} \\ \text{contribution}\end{array}\right)$$

Substitute this into Eq (1):

$$\hat{x}_k = \phi_{k-1}\hat{x}_{k-1} + \delta x + K_k(z_k - H_k(\phi_{k-1}\hat{x}_{k-1} + \delta x))$$

Now, approximate ϕ as $I + F\Delta t$ and let $K_k = K\Delta t$ and retain only 1st-order terms in Δt.

This leads to: (note that δx is of the order of Δt)

$$\hat{x}_k = \hat{x}_{k-1} + F\Delta t\, \hat{x}_{k-1} + \delta x + K\Delta t[z_k - H\hat{x}_{k-1}] \qquad (2)$$

Now note that the deterministic contribution δx can be written as $Bu\Delta t$. (See Eq 5.35 and note that the input is Bu in this case.) Substitute this into Eq(2) and rewrite as

$$\frac{\hat{x}_k - \hat{x}_{k-1}}{\Delta t} = F\hat{x}_{k-1} + K[z_k - H\hat{x}_{k-1}] + Bu$$

Now, let $\Delta t \to 0$ and note that $x_k \to x_{k-1}$ as $\Delta t \to 0$. Thus

$$\dot{\hat{x}} = F\hat{x} + K[z - H\hat{x}] + Bu$$

CHAPTER 8

8.1 Take advantage of the work in Example 4.3 and begin with

$$\frac{S_{s+m,s}}{S_{stm}^{-}} = \frac{2}{(-S+\sqrt{3})(S+1)} = \frac{\sqrt{3}-1}{S+1} + \frac{\sqrt{3}-1}{-S+\sqrt{3}}$$

Sketches of inverse and shifted inverse:

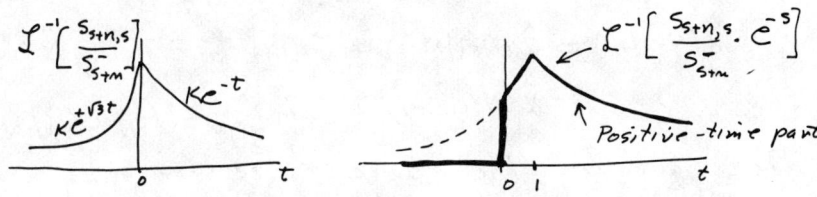

$$\mathcal{L}^{-1}\left[\frac{S_{s+m,s}}{S_{stm}^{-}}\right] \qquad Ke^{-t} \qquad Ke^{+\sqrt{3}t} \qquad \mathcal{L}^{-1}\left[\frac{S_{s+m,s}}{S_{stm}^{-}} \cdot e^{-S}\right] \qquad \text{Positive -time part}$$

We need to find the Laplace transform of the positive-time part shown above. Call pos. time part $f(t)$. Then

$$f(t) = (\sqrt{3}-1)e^{\sqrt{3}(t-1)}[u(t) - u(t-1)] + (\sqrt{3}-1)e^{-(t-1)}u(t-1)$$

$$\mathcal{L}[f(t)] = \frac{(\sqrt{3}-1)\left[(e^{-\sqrt{3}}-e^{-S})(S+1) + e^{-S}(S-\sqrt{3})\right]}{(S-\sqrt{3})(S+1)}$$

Optimal transfer function is then

$$G(s) = \frac{1}{S_{stm}^{+}}\left\{\mathcal{L}[f(t)]\right\}$$

$$= \frac{S+1}{S+\sqrt{3}} \cdot \frac{(\sqrt{3}-1)\left[(e^{-\sqrt{3}}-e^{-S})(S+1) + e^{-S}(S-\sqrt{3})\right]}{(S-\sqrt{3})(S+1)}$$

(At first glance, the above result may appear to be noncausal because of the $S-\sqrt{3}$ term in the denominator. However, it is causal and note it is irrational in form because of the e^{-S} terms.)

8.2 (a) The linear connection between the batched measurement and x_{N-1} is given in the problem statement. Begin with $\hat{x}(N-1/N-2)$ and $P(N-1/N-2)$ and update estimate with 3_{N-1}:

$$\hat{x}(N-1/N-1) = \hat{x}(N-1/N-2) + K_{N-1}\left(3_{N-1} - H_{N-1}\hat{x}(N-1/N-2)\right)$$

$$P(N-1/N-1) = \text{Usual P-update equation}$$

Now update estimate again using 3_N:

$$\hat{x}(N-1/N) = \hat{x}(N-1/N-1) + \text{Gain}\left[3_N - H_N\phi_{N-1}\hat{x}(N-1/N-1)\right]$$

$$= \hat{x}(N-1/N-1) + \text{Gain}\left[3_N - H_N\hat{x}(N/N-1)\right] \quad (1)$$

The "Gain" in the above expression may be written as

$$\text{Gain} = P(N-1/N-1)(H_N\phi_{N-1})^T\left[(H_N\phi_{N-1})P(N-1/N-1)(H_N\phi_{N-1})^T\right.$$
$$\left. + (H_N Q_N H_N^T + R_N)\right]^{-1}$$

$$= P(N-1/N-1)\phi_{N-1}^T H_N^T\left[H_N\left(\phi_{N-1}P(N-1/N-1)\phi_{N-1}^T + Q_N\right)H_N^T\right.$$
$$\left. + R_N\right]^{-1}$$

Now insert $\overset{-1}{P}(N/N-1)\cdot P(N/N-1)$

and note $\left[\phi_{N-1}P(N-1/N-1)\phi_{N-1}^T + Q_N\right] = P(N/N-1)$. This yields

$$\text{Gain} = P(N-1/N-1)\phi_{N-1}^T \overset{-1}{P}(N/N-1) P(N/N-1) H_N^T$$
$$\left[H_N P(N/N-1) H_N^T + R_N\right]^{-1}$$

$$= \left[P(N-1/N-1)\phi_{N-1}^T \overset{-1}{P}(N/N-1)\right] K_N \quad (2)$$

Now compare Eqs. (1), (2) above with Rauch formula:

$$\hat{x}(N-1/N) = \hat{x}(N-1/N-1) + A(N-1)\left[\hat{x}(N/N) - \hat{x}(N/N-1)\right]$$

But $\hat{x}(N/N) = \hat{x}(N/N-1) + K_N\left[3_N - H_N\hat{x}(N/N-1)\right]$

8.2 (cont.)

Substituting for $\hat{x}(N/N)$ in Rauch formula yields

$$\hat{x}(N-1/N) = \hat{x}(N-1/N-1) + A(N-1) K_N [3_N - H_N \hat{x}(N/N-1)] \quad (3)$$

we can now see that this is the equivalent of Eqs. (1) and (2) if we let

$$A(N-1) = P(N-1/N-1) \phi_{N-1}^T P^{-1}(N/N-1)$$

(b) To generalize, let k be interior point where estimate is desired. Batch together $3_{k+1}, 3_{k+2} \cdots 3_N$ and call this vector y_{k+1}. Consider the measurement relationship at t_k to be:

$$\begin{bmatrix} 3_k \\ \cdots \\ y_{k+1} \end{bmatrix} = \begin{bmatrix} H_k \\ \cdots \\ M_{k+1}\phi_k \end{bmatrix} x_k + \begin{bmatrix} --- v_k --- \\ M_{k+1} w_k + u_{k+1} \end{bmatrix} \quad (4)$$

The lower partitioned part of Eq (4) comes from

$$y_{k+1} = M_{k+1} x_{k+1} + u_{k+1}$$
$$= M_{k+1}(\phi_k x_k + w_k) + u_{k+1}$$
$$= (M_{k+1}\phi_k) x_k + (M_{k+1} w_k + u_{k+1})$$

We can now argue that $(M_{k+1} w_k + u_{k+1})$ is not correlated with v_k because u_{k+1} only involves w's and not the v's. Therefore 3_k and y_{k+1} may be assimilated in 2 steps just as in part (a). First, write $\hat{x}(k/k)$ as

$$\hat{x}(k/k) = \hat{x}(k/k-1) + K_k(3_k - H_k \hat{x}(k/k-1))$$

Next, assimilate y_{k+1} to obtain $\hat{x}(k/N)$

$$\hat{x}(k/N) = \hat{x}(k/k) + Gain[y_{k+1} - M_{k+1}\phi_k \hat{x}(k/k)]$$

From here on, the derivation is similar to part (a) except for notation.

8.3

The model:

$$\dot{x} = -x + w(t)$$

$w \longrightarrow \boxed{\dfrac{\sqrt{2}}{s+1}} \longrightarrow x$

(unity white noise)

At $t = 1$ sec

$\therefore \quad \phi = e^{-1}, \quad Q = 1 - e^{-2}, \quad R = 1, \quad H = 1$

$\hat{x}_0^- = 0, \quad P_0^- = 1$

(a) The forward sweep using the regular Kalman filter equations leads to

$$\hat{x}(1|1) = \frac{e^{-1}}{4 - e^{-2}} \, z_0 + \frac{2 - e^{-2}}{4 - e^{-2}} \, z_1$$

The backward sweep consists of just one step.

$$\hat{x}(0|1) = \hat{x}(0|0) + A(0)\left[\hat{x}(1|1) - \hat{x}(1|0)\right]$$

where

$$A(0) = P(0|0) \cdot e^{-1} \cdot \frac{1}{P(1|0)} = \frac{e^{-1}}{2 - e^{-2}}$$

The final result:

$$\hat{x}(0|1) = \frac{2 - e^{-2}}{4 - e^{-2}} \, z_0 + \frac{e^{-1}}{4 - e^{-2}} \, z_1$$

(b) Fraser - Potter scheme:

The forward filter is obtained by assimilating z_0 only at $t = 0$. The results of this are:

$$\hat{x}_{for.} = \frac{1}{2} \, z_0 \quad ; \quad P_{for.} = \frac{1}{2}$$

The backward filter is obtained by beginning at $t = 1$ with $\hat{x}(1)_{back.} = 0$ and $P(1)_{back.} = \infty$.
Now update using z_1. This leads to:

$$\hat{x}(1)_{back.} = z_1 \quad , \quad P(1)_{back.} = 1$$

126

8.3 (cont.)

Now, project backward to $t=0$.

$$\hat{x}(0)_{back.} = e^{-1} \cdot \hat{x}(1)_{back} = e^{-1} z_1$$

$$\bar{P}(0)_{back.} = \phi^{-1}(P(1)_{back} + Q)\phi^{-1}$$
$$= e[1 + (1-e^{-2})]e = 2e^2 - 1$$

Finally, combine the 2 estimates:

$$\text{weight factor for } \hat{x}_{for} = \frac{2e^2-1}{(2e^2-1)+\frac{1}{2}} = \frac{4e^2-2}{4e^2-1}$$

$$\text{weight factor for } \hat{x}_{back} = \frac{\frac{1}{2}}{(2e^2-1)+\frac{1}{2}} = \frac{1}{4e^2-1}$$

Final estimate:

$$\hat{x}(0|1) = \frac{2e^2-1}{4e^2-1} z_0 + \frac{e}{4e^2-1} z_1 \qquad \begin{array}{l}\text{(same as for}\\ \text{RTS Algorithm)}\end{array}$$

(c) Linear equations for weight factors k_0 and k_1 are:

$$\begin{bmatrix} 2 & e^{-1} \\ e^{-1} & 2 \end{bmatrix}\begin{bmatrix} k_0 \\ k_1 \end{bmatrix} = \begin{bmatrix} 1 \\ e^{-1} \end{bmatrix}$$

Solving for k_0 and k_1 yields

$$k_0 = \frac{2-e^{-2}}{4-e^{-2}} \quad ; \quad k_1 = \frac{e^{-1}}{4-e^{-2}}$$

Optimal estimate is then:

$$\hat{x}(0|1) = \frac{2-e^{-2}}{4-e^{-2}} z_0 + \frac{e^{-1}}{4-e^{-2}} z_1 \qquad \begin{array}{l}\text{(same as}\\ \text{obtained}\\ \text{in parts}\\ \text{(a) and (b).)}\end{array}$$

(d) For the fixed-point algorithm, let $k=0$ and $j=1$. Then

$$\hat{x}(0|1) = \hat{x}(0|0) + B(1)[\hat{x}(1|1) - \hat{x}(1|0)]$$

where $B(1) = \Pi A(0) = A(0)$

This is identical with RTS algorithm. See part (a).

127

<u>8.4</u>   Begin with the gain expression Eq. (8.3.2)
$$A(k) = P(k/k)\,\phi^T(k+1,k)\,\bar{P}^{-1}(k+1/k)$$
Or, postmultiplying by $P(k+1/k)$ yields
$$A(k)\,P(k+1/k) = P(k/k)\,\phi^T(k+1,k)$$
Now approximate $P(k+1/k)$ and $\phi^T(k+1,k)$ as
$$P(k+1/k) \approx \phi\,P(k/k)\,\phi^T + GQG^T\Delta t, \text{ and } \phi^T \approx I + F^T\Delta t$$
Next, shorten $P(k/k)$ to just P and substitute.
$$A(k)\left[(I+F\Delta t)P(I+F^T\Delta t) + GQG^T\Delta t\right] = P[I + F^T\Delta t]$$
Now, neglect higher order terms in Δt and rewrite:
$$(A-I)P = PF^T\Delta t - AFP\Delta t - APF^T\Delta t - AGQG^T\Delta t$$
Note $A \to I$ as $\Delta t \to 0$, so $PF^T\Delta t \approx APF^T\Delta t$.
Cancelling terms and rearranging leads to
$$A \approx I - (F + GQG^T P^{-1})\Delta t$$
Now, substitute this in estimate equation, Eq.(8.3.1)
$$\hat{x}(k/N) = \hat{x}(k/k) + \left[I - (F+GQG^T P^{-1})\Delta t\right]\left[\hat{x}(k+1/N) - \hat{x}(k+1/k)\right]$$
Rearrange terms:
$$\hat{x}(k/N) - \hat{x}(k+1/N) = \hat{x}(k/k) - \hat{x}(k+1/k)$$
$$- (F+GQG^T P^{-1})\Delta t\left[\hat{x}(k+1/N) - \hat{x}(k+1/k)\right]$$
Next, note that $\hat{x}(k+1/k)$ can be written as
$$\hat{x}(k+1/k) = \phi\,x(k/k) \approx (I+F\Delta t)x(k/k)$$
and
$$\hat{x}(k/k) - \hat{x}(k+1/k) \approx -F\Delta t\,x(k/k)$$

Now, divide by Δt and pass to the limit and note that $\hat{x}(k+1/k) \to \hat{x}(k/k)$ (or simply $\hat{x}(t/t)$).
Final result:
$$\dot{\hat{x}}(t/T) = F\hat{x}(t/T) + GQG^T P^{-1}\left[\hat{x}(t/T) - \hat{x}(t/t)\right]$$

8.4 (cont.)

For the error covariance, begin by writing Eq. (8.3.3) in the form:

$$P(k/N) = P(k/k) + A\,P(k+1/N)A^T - A\,P(k+1/k)A^T$$

Now approximate A as before and obtain

$$P(k/N) = P(k/k) + [I - (\ )\Delta t]\,P(k+1/N)[I - (\ )\Delta t]^T$$
$$- [I - (\ )\Delta t]\,P(k+1/k)[I - (\ )\Delta t]^T$$

where "$(\ )$" is $(F + GQG^T P^{-1})$. Now neglect higher order terms in Δt and rearrange:

$$P(k/N) - P(k+1/N) = P(k/k) - P(k+1/k)$$
$$+ \Big[-(\ )P(k+1/N) - P(k+1/N)(\ )^T$$
$$+ (\ )P(k+1/k) + P(k+1/k)(\ )^T\Big]\Delta t$$

Now note

$$P(k+1/k) = \phi\,P(k/k)\phi^T + GQG^T\Delta t$$
$$\approx P(k/k) + FP(k/k)\Delta t + P(k/k)F^T\Delta t + GQG^T\Delta t$$

Or $P(k+1/k) - P(k/k) \approx [\quad 3\ terms \quad]\Delta t$

Right side of $P(k/N) - P(k+1/N)$ expression is now first order in Δt and we can divide by Δt. Note $P(k+1/k) \to P(k/k)$ in the limit, so $FP(k/k)$ cancels $FP(k+1/k)$, and $P(k/k)F^T$ cancels $P(k+1/k)F^T$. The final result is

$$\dot P(t/T) = [F + GQG^T P^{-1}(t/t)]\,P(t/T)$$
$$+ P(t/T)[F + GQG^T P^{-1}(t/t)]^T - GQG^T$$

129

<u>8.5</u>   Diff. eq. for $P(t|T)$ is
$$\dot{P}(t|T) = \left[F + GQG^T\dot{P}'(t|t)\right]P(t|T) + P(t|T)\left[F + GQG^T\dot{P}'(t|t)\right]$$
$$- GQG^T$$
The model parameters are:
$$F=0,\ G=1,\ Q=q,\ R=r,\ H=1$$
(a) First solve the filter equation for steady-state.
$$\dot{P} = FP + PF^T - PH^TR^{-1}HP + GQG$$
$$\dot{P} = 0 + 0 - P^2/r + q$$
Now let $\dot{P}=0$ and solve for P (steady-state).
$$P = \sqrt{rq} \triangleq \alpha$$
we now let this be the $P(t|t)$ in smoothing eq.

$$\dot{P}(t|T) = \frac{q}{\sqrt{rq}}\ P(t|T) + P(t|T)\frac{q}{\sqrt{rq}} - q;\quad P(T|T) = \alpha$$

or $\dot{P}(t|T) = 2\sqrt{\frac{q}{r}}\ P(t|T) - q$;   Bound. Cond: $P(T|T) = \alpha$

It can now be easily verified that
$$P(t|T) = \frac{\alpha}{2}\left[1 + e^{-2\beta(T-t)}\right]$$
is the correct solution.
Sketch:

Approx. sol. for $P(t|t)$

(b) We assumed $P(t|t)$ to be a constant (i.e., α) in the smoothing diff. eq. This is a fairly good approximation for t near T, but not for t near zero. Therefore, we should not expect our "solution" to be valid near the origin.

8.6   First write the final 2 measurements as

$$z_N = x_{1N} + x_{2N} + v_N$$
$$z_{N-1} = x_{1(N-1)} + x_{2(N-1)} + v_{N-1}$$

However,

$$x_{1N} = e^{-1} x_{1(N-1)} + q_1$$
$$x_{2N} = e^{-2} x_{2(N-1)} + q_2$$

q_1 and q_2 are normal, zero-mean and independent. Let their σ's be σ_1^2 and σ_2^2

Now, substitute and batch together z_N and z_{N-1}:

$$\begin{bmatrix} z_N \\ z_{N-1} \end{bmatrix} = \begin{bmatrix} e^{-1} & e^{-2} \\ 1 & 1 \end{bmatrix} \begin{bmatrix} x_{1(N-1)} \\ x_{2(N-1)} \end{bmatrix} + \begin{bmatrix} q_1 + q_2 + v_N \\ v_{N-1} \end{bmatrix} \quad (1)$$

Now, solve for $x_{1(N-1)}$ and $x_{2(N-1)}$ just as if measurements were noiseless.

$$\hat{x}_{1(N-1)} = \frac{\begin{vmatrix} z_N & e^{-2} \\ z_{N-1} & 1 \end{vmatrix}}{\begin{vmatrix} e^{-1} & e^{-2} \\ 1 & 1 \end{vmatrix}}, \text{ and } \hat{x}_{2(N-1)} = \frac{\begin{vmatrix} e^{-1} & z_N \\ 1 & z_{N-1} \end{vmatrix}}{\begin{vmatrix} e^{-1} & e^{-2} \\ 1 & 1 \end{vmatrix}}$$

Or

$$\hat{x}_{1(N-1)} = \frac{1}{1-e} z_{N-1} - \frac{e^2}{1-e} z_N$$
$$\hat{x}_{2(N-1)} = -\frac{e}{1-e} z_{N-1} + \frac{e^2}{1-e} z_N$$

"Noiseless" estimates at $t = N-1$

Next, take the longer more rigorous approach and process z_N and z_{N-1} sequentially in a backward filter. We begin at $t = N$ with ($*$ means a priori)

$$P_N^* = \begin{bmatrix} M & 0 \\ 0 & M \end{bmatrix}, \quad \hat{x}_N^* = \begin{bmatrix} 0 \\ 0 \end{bmatrix}$$

where M is very large (eventually, we let $M \to \infty$).

8.6 (cont.)

First assimilate z_N. This leads to

$$\hat{x}_{N(Back)} = \begin{bmatrix} \frac{M}{2M+r} \\ \frac{M}{2M+r} \end{bmatrix} z_N, \text{ and } P_{N(Back)} = \begin{bmatrix} \frac{M(M+r)}{2M+r} & \frac{-M^2}{2M+r} \\ \frac{-M^2}{2M+r} & \frac{M(M+r)}{2M+r} \end{bmatrix}$$

Next, project back one step using
Eqs. (8.6.25) and (8.6.26). This leads to:

$$\hat{x}_{N-1}^* = \begin{bmatrix} \frac{1}{2} e\, z_N \\ \frac{1}{2} e^2 z_N \end{bmatrix} \quad (\text{For large } M)$$

And

$$P_{N-1}^* = \begin{bmatrix} \left(\frac{M(M+r)}{2M+r} + \sigma_1^2\right) e^2 & \left(\frac{-M^2}{2M+r}\right) e^3 \\ \left(\frac{-M^2}{2M+r}\right) e^3 & \left(\frac{M(M+r)}{2M+r} + \sigma_2^2\right) e^4 \end{bmatrix}$$

Now assimilate z_{N-1} using usual Kalman gain
formula. In computing gain, let $M \to \infty$. This yields

$$(Gain)_{N-1} = \begin{bmatrix} \frac{1}{1-e} \\ \frac{-e}{1-e} \end{bmatrix}$$

The updated estimate is then

$$\hat{x}_{N-1(Back)} = \begin{bmatrix} \frac{1}{2} e\, z_N \\ \frac{1}{2} e^2 z_{N-1} \end{bmatrix} + \begin{bmatrix} \frac{1}{1-e} \\ \frac{-e}{1-e} \end{bmatrix} \left[z_{N-1} - \left(\frac{1}{2} e\, z_N + \frac{1}{2} e^2 z_N \right) \right]$$

$$= \begin{bmatrix} \frac{1}{1-e} z_{N-1} - \frac{e^2}{1-e} z_N \\ \frac{-e}{1-e} z_{N-1} + \frac{e^2}{1-e} z_N \end{bmatrix}$$

This is identical to the deterministic (i.e., noiseless)
result.

Both the noiseless and backward-filter
estimates are identical functions of z_N and z_{N-1}.
Thus they must have identical error covariances.

<u>8.6 (con't.)</u>   The easiest way to find the error covariance is to go back to Eq. (1). First denote $\begin{bmatrix} e_1^{-1} & e_1^{-2} \end{bmatrix}$ as matrix A. Then

$$\begin{bmatrix} x_1 \\ x_2 \end{bmatrix}_{N-1} = A^{-1}\begin{bmatrix} 3_N \\ 3_{N-1} \end{bmatrix} - A^{-1}\begin{bmatrix} q_1 + q_2 + v_N \\ v_{N-1} \end{bmatrix}$$

But our estimate is $A^{-1}\begin{bmatrix} 3_N \\ 3_{N-1} \end{bmatrix}$. Thus the mean square error is

$$E[(\text{error})(\text{error})^T] = A^{-1}E\left\{ \begin{pmatrix} q_1 + q_2 + v_N \\ v_{N-1} \end{pmatrix} \begin{pmatrix} q_1 + q_2 + v_N & v_{N-1} \end{pmatrix} \right\} A^{-1^T}$$

Now. note that q_1, q_2, v_N, and v_{N-1} are uncorrelated. The expectation is then

$$\begin{bmatrix} E(q_1^2) + E(q_2^2) + E(v_N^2) & 0 \\ 0 & E(v_{N-1}^2) \end{bmatrix} = \begin{bmatrix} \sigma_1^2 + \sigma_2^2 + r & 0 \\ 0 & r \end{bmatrix}$$

And the error covariance is

$$\text{Error Cov.} = A^{-1}\begin{bmatrix} \sigma_1^2 + \sigma_2^2 + r & 0 \\ 0 & r \end{bmatrix} A^{-1^T}$$

It is now routine to evaluate this. Thus, the backward filter can be started at $t = N-1$ with the "noiseless" estimate and the above error covariance matrix.

<u>8.7</u>   After $j+1$ steps the usual forward filter has assimilated $j+1$ measurements "behind" the present point of estimation. On the other hand, the fixed point filter has processed $j+1$ measurements "ahead" of its point of estimation. Because of the even symmetry of the auto correlation function, one would expect both estimates to be of the same quality.

9.1 Nonlinear dynamical equations are:

$$\ddot{r} - r\dot{\theta}^2 + K/r^2 = W_r(t)$$
$$r\ddot{\theta} + 2\dot{r}\dot{\theta} = W_\theta(t)$$

Let $x_1 = r$, $x_2 = \dot{r}$, $x_3 = \theta$, $x_4 = \dot{\theta}$. Then state eqs. are:

$$
\begin{bmatrix} \dot{x}_1 \\ \dot{x}_2 \\ \dot{x}_3 \\ \dot{x}_4 \end{bmatrix}
=
\begin{bmatrix} x_2 \\ x_1 x_4^2 - K/x_1^2 \\ x_4 \\ -2x_2 x_4/x_1 \end{bmatrix}
+
\begin{bmatrix} 0 \\ W_r \\ 0 \\ W_\theta' \end{bmatrix}
, \quad W_\theta' \approx \frac{W_\theta}{R_0}
$$

Next, find $\partial f/\partial \underline{x}$ as defined by Eq. (9.1.8)

$$
\frac{\partial f}{\partial \underline{x}} =
\begin{bmatrix}
0 & 1 & 0 & 0 \\
\left(x_4^2 + \dfrac{2K}{x_1^3}\right) & 0 & 0 & 2x_1 x_4 \\
0 & 0 & 0 & 1 \\
\dfrac{2x_2 x_4}{x_1^2} & -2x_4/x_1 & 0 & -\dfrac{2x_2}{x_1}
\end{bmatrix}
$$

$$
=
\begin{bmatrix}
0 & 1 & 0 & 0 \\
\dot{\theta}^2 + 2\dfrac{K}{r^3} & 0 & 0 & 2r\dot{\theta} \\
0 & 0 & 0 & 1 \\
\dfrac{2\dot{r}\dot{\theta}}{r^2} & -2\dot{\theta}/r & 0 & -2\dot{r}/r
\end{bmatrix}
$$

Now evaluate above along nominal trajectory where $\dot{\theta} = w_0$, $w_0^2 = K/R_0^3$, $r = R_0$, $\dot{r} = 0$. This yields

$$
\left.\frac{\partial f}{\partial \underline{x}}\right|_{\underline{x} = \underline{x}^*}
=
\begin{bmatrix}
0 & 1 & 0 & 0 \\
3w_0^2 & 0 & 0 & 2R_0 w_0 \\
0 & 0 & 0 & 1 \\
0 & -2w_0/R_0 & 0 & 0
\end{bmatrix}
$$

<u>9.2</u>   Dynamical equation for "up" trajectory is

$$\ddot{y} = -g - \frac{D}{m}\dot{y}^2$$

The state model:
Let $y = x_1$, $\dot{y} = x_2$. Then

$$\begin{bmatrix} \dot{x_1} \\ \dot{x_2} \end{bmatrix} = \begin{bmatrix} x_2 \\ -g - \frac{D}{m}x_2^2 \end{bmatrix} + \begin{bmatrix} 0 \\ 0 \end{bmatrix}, \quad \underline{x}(0) = \begin{bmatrix} 0 \\ 85 \end{bmatrix}$$

The measurement relationship:

$$z = x_1 + v, \qquad v \text{ is uncorrelated error}$$

In the nonlinear model, $\underline{f}$ and $\underline{h}$ are

$$\underline{f} = \begin{bmatrix} x_2 \\ -g - \frac{D}{m}x_2^2 \end{bmatrix} ; \quad \underline{h} = [x_1]$$

Next, use Eq. (9.1.8) to obtain the linearized F
and H matrices.

$$\frac{\partial \underline{f}}{\partial \underline{x}} = \begin{bmatrix} 0 & 1 \\ 0 & -\frac{2D}{m}x_2 \end{bmatrix} ; \quad \frac{\partial \underline{h}}{\partial \underline{x}} = [1 \quad 0]$$

First, note the H matrix is constant so this
presents no problem. However, $\partial \underline{f}/\partial \underline{x}$ involves x_2,
so this must be evaluated along a particular
deterministic trajectory. Find trajectory next.
        Let $\dot{y}$ be velocity variable v, and rewrite
diff. eq. as $\dot{v} = -g - \frac{D}{m}v^2$

Or $\dfrac{dv}{v^2 + \frac{mg}{D}} = -\frac{D}{m}dt$ (1)

9.2 (Cont.)

Integrate both sides of (1) to get

$$\frac{1}{\sqrt{\frac{mg}{D}}} \tan^{-1}\left(\frac{v}{\sqrt{\frac{mg}{D}}}\right) = -\frac{D}{m}t + (\text{Constant})$$

Now use boundary condition, $v = V_0$ at $t=0$, and evaluate the constant. This leads to

$$v = \sqrt{\frac{mg}{D}}\ \tan\left(-\sqrt{\frac{Dg}{m}}\ t + C'\right)$$

where

$$C' = \tan^{-1}\left(V_0/\sqrt{\frac{mg}{D}}\right)$$

Using the numerical values given in the problem:

$$v = 59.18\ \tan\left(-.1656\, t + .9626\right)$$

This can now be integrated to get position. The result is:

$$y = \frac{m}{D}\left[\ln \frac{\cos\left(-\sqrt{\frac{Dg}{m}}\, t + C'\right)}{\cos C'}\right]$$

or

$$y = 357.14\left[\ln \frac{\cos\left(-.1656\, t + .9626\right)}{.5714}\right]$$

(Maximum height is about 200 m. and this occurs after an elapsed time of about 5.81 sec.)

The linearized model:

Initial Conditions:

$$\hat{x}_0 = \begin{bmatrix} 0 \\ 0 \end{bmatrix}, \quad P_0^- = \begin{bmatrix} 0 & 0 \\ 0 & 1 \end{bmatrix}$$

(Note that the states are perturbations from the nominal trajectory.)

$$H = \begin{bmatrix} 1 & 0 \end{bmatrix}, \quad (\text{Same as in nonlinear model})$$

$$Q = \begin{bmatrix} 0 & 0 \\ 0 & 0 \end{bmatrix}, \quad (\text{The only randomness is in the initial velocity and measurement.})$$

136

9.2 (con't)

$$R = [.25]$$

Transition matrix: Strictly speaking, we should carefully integrate (say, numerically) the linearized differential equation

$$\begin{bmatrix} \dot{x}_1 \\ \dot{x}_2 \end{bmatrix} = \begin{bmatrix} 0 & 1 \\ 0 & -\frac{2D}{m}x_2 \end{bmatrix} \begin{bmatrix} x_1 \\ x_2 \end{bmatrix}$$

as discussed in Sec. 5.3 to get ϕ_k for each step. However, the step size is reasonably small, so it would seem reasonable to approximate the $-\frac{2D}{m}x_2$ term as constant over the interval. The value of the constant should be determined from the nominal trajectory rather than the Kalman filter estimate. Otherwise, we would have an extended rather than a simple linear-ized Kalman filter. For example, for the first step:

$$v_{nominal}(0) = 85 \text{ ft/sec}$$

$$v_{nominal}(.1) = 59.18 \tan(-.016565 + .9626)$$
$$\approx 82.07 \text{ ft/sec}$$

We might well approximate x_2 as 83.5 ft/sec. This would lead to a ϕ_k for the first interval:

$$\phi_k(\text{First step}) = \begin{bmatrix} 1 & \frac{1}{a}(1-e^{-a\Delta t}) \\ 0 & e^{-a\Delta t} \end{bmatrix}$$

where $\Delta t = .1$ and $a = \frac{2D}{m} \cdot 83.5$. Repeat this for each step. Remember, we will be estimating the perturbations from the nominal. Total is the nominal plus the perturbation.

137

9.3 (a) The simplest way to justify that the residuals are least squares is to assume all processes are zero-mean and Gaussian. Then use the following argument

(1) $\hat{x}_{\bar{k}}$ is the conditional mean of x_k, conditioned on $z_0, z_1, \cdots, z_{k-1}$.

(2) A linear transformation on $\hat{x}_{\bar{k}}$ also leads to an optimal estimate, so $H_k\hat{x}_{\bar{k}}$ is the cond. mean of random variable z_k, conditioned on $z_0, z_1, \cdots, z_{k-1}$.

(3) In the Gaussian case we know that the conditional mean is also minimum-mean-square. Thus the residual, which by definition is the estimation error in estimating z_k given $z_0, z_1, \cdots z_k$, must be a minimum in a mean square sense.

(b) To show the sequence of residuals is white, use the same Gaussian assumption as in part (a). Consider the sequence:

$$z_0 - H_0\hat{x}_{\bar{0}}, \; z_1 - H_1\hat{x}_{\bar{1}}, \; \cdots, \; z_k - H_k\hat{x}_{\bar{k}}$$

Call the last term the estimation error $\bar{e}_k$. Now, the orthogonality principle states that this error is orthogonal to _all_ previous measurements (i.e., $z_0, z_1, z_2; \cdots z_{k-1}$). Therefore,

$$E\left\{\bar{e}_k \left(z_{k-1} - H_{k-1}\hat{x}_{\bar{k-1}}\right)\right\} = E\left[\bar{e}_k \left(-H_{k-1}\hat{x}_{\bar{k-1}}\right)\right]$$

But, $\hat{x}_{\bar{k-1}}$ is formed as a linear combination of measurements $z_0, z_1, \cdots z_{k-2}$. Thus the above expectation is zero. This argument can now be repeated for $\left(z_{k-2} - H_{k-2}\hat{x}_{\bar{k-2}}\right)$ etc.

APPENDIX

Exercise A.1

(a) $\mathcal{L}^{-1}[F(s)] = \frac{1}{2\pi}\int_{-\infty}^{\infty} \text{Re}\left[\frac{e^{j\omega t}}{j\omega+2}\right]d\omega + j\frac{1}{2\pi}\int_{-\infty}^{\infty}\text{Im}\left[\frac{e^{j\omega t}}{j\omega+2}\right]d\omega$

Clearly, $\text{Im}[\ ]$ is an odd function of ω and thus the second term is zero. Rewriting the first term leads to

$$\mathcal{L}^{-1}[F(s)] = \frac{1}{2\pi}\int_{-\infty}^{\infty}\frac{2\cos\omega t}{\omega^2+4}\,d\omega + \frac{1}{2\pi}\int_{-\infty}^{\infty}\frac{\omega\sin\omega t}{\omega^2+4}\,d\omega$$

Now let $\omega = 2x$ and note even symmetry in first term. Thus

$$\frac{1}{2\pi}\int_{-\infty}^{\infty}\frac{2\cos\omega t}{\omega^2+4}\,d\omega = \frac{1}{\pi}\int_{0}^{\infty}\frac{\cos(2t)x}{x^2+1}\,dx = \frac{1}{2}e^{-|2t|} \quad \text{(From integral tables)}$$

The second term may be written as

$$\frac{1}{2\pi}\int_{-\infty}^{\infty}\frac{\omega\sin\omega t}{\omega^2+4}\,d\omega = \frac{1}{2\pi}\int_{-\infty}^{\infty}\frac{d}{dt}\left[-\frac{\cos\omega t}{\omega^2+4}\right]d\omega = -\frac{d}{dt}\left[\frac{1}{2\pi}\int_{-\infty}^{\infty}\frac{\cos\omega t}{\omega^2+4}\,d\omega\right]$$

$$= -\frac{d}{dt}\left[\frac{1}{4}e^{-|2t|}\right] = \begin{cases}\frac{1}{2}e^{-2t}, & t>0 \\ -\frac{1}{2}e^{2t}, & t<0\end{cases}$$

$$\therefore \mathcal{L}^{-1}[F(s)] = [\text{Sum of above terms}] = \begin{cases}e^{-2t}, & t>0 \\ 0, & t<0\end{cases}$$

(b) When $t=0$:

$$\mathcal{L}^{-1}\{F(s)\} = \frac{1}{2\pi}\int_{-\infty}^{\infty}\frac{2}{\omega^2+4}\,d\omega = \frac{1}{2}$$

This is the midpoint between the two extremes at the point of discontinuity, just as occurs in a Fourier series.